# STUDENT WORKBOOK
## WITH MODERN PHYSICS

# PHYSICS
### FOR SCIENTISTS AND ENGINEERS

#### A STRATEGIC APPROACH

Randall D. Knight
California Polytechnic State University
San Luis Obispo

PEARSON

Addison
Wesley

San Francisco    Boston    New York
Capetown    Hong Kong    London    Madrid    Mexico City
Montreal    Munich    Paris    Singapore    Sydney    Tokyo    Toronto

Cover Credit: Rainbow/PictureQuest

ISBN: 0-8053-8961-X

# Table of Contents

**Part I**  Newton's Laws

| | | |
|---|---|---|
| **Chapter 1** | Concepts of Motion | 1-1 |
| **Chapter 2** | Kinematics: the Mathematics of Motion | 2-1 |
| **Chapter 3** | Vectors and Coordinate Systems | 3-1 |
| **Chapter 4** | Force and Motion | 4-1 |
| **Chapter 5** | Dynamics I: Motion Along a Line | 5-1 |
| **Chapter 6** | Dynamics II: Motion in a Plane | 6-1 |
| **Chapter 7** | Dynamics III: Motion in a Circle | 7-1 |
| **Chapter 8** | Newton's Third Law | 8-1 |

**Part II**  Conservation Laws

| | | |
|---|---|---|
| **Chapter 9** | Impulse and Momentum | 9-1 |
| **Chapter 10** | Energy | 10-1 |
| **Chapter 11** | Work | 11-1 |

**Part III**  Applications of Newtonian Mechanics

| | | |
|---|---|---|
| **Chapter 12** | Newton's Theory of Gravity | 12-1 |
| **Chapter 13** | Rotation of a Rigid Body | 13-1 |
| **Chapter 14** | Oscillations | 14-1 |
| **Chapter 15** | Fluids and Elasticity | 15-1 |

**Part IV**  Thermodynamics

| | | |
|---|---|---|
| **Chapter 16** | A Macroscopic Description of Matter | 16-1 |
| **Chapter 17** | Work, Heat, and the First Law of Thermodynamics | 17-1 |
| **Chapter 18** | The Micro/Macro Connection | 18-1 |
| **Chapter 19** | Heat Engines and Refrigerators | 19-1 |

**Part V**  Waves and Optics

| | | |
|---|---|---|
| **Chapter 20** | Traveling Waves | 20-1 |
| **Chapter 21** | Superposition | 21-1 |
| **Chapter 22** | Wave Optics | 22-1 |
| **Chapter 23** | Ray Optics | 23-1 |
| **Chapter 24** | Modern Optics and Matter Waves | 24-1 |

| Part VI | Electricity and Magnetism | |
|---|---|---|
| Chapter 25 | Electric Charges and Forces | 25-1 |
| Chapter 26 | The Electric Field | 26-1 |
| Chapter 27 | Gauss's Law | 27-1 |
| Chapter 28 | Current and Conductivity | 28-1 |
| Chapter 29 | The Electric Potential | 29-1 |
| Chapter 30 | Potential and Field | 30-1 |
| Chapter 31 | Fundamentals of Circuits | 31-1 |
| Chapter 32 | The Magnetic Field | 32-1 |
| Chapter 33 | Electromagnetic Induction | 33-1 |
| Chapter 34 | Electromagnetic Fields and Waves | 34-1 |
| Chapter 35 | AC Circuits | 35-1 |

| Part VII | Relativity and Quantum Physics | |
|---|---|---|
| Chapter 36 | Relativity | 36-1 |
| Chapter 37 | The End of Classical Physics | 37-1 |
| Chapter 38 | Quantization | 38-1 |
| Chapter 39 | Wave Functions and Probabilities | 39-1 |
| Chapter 40 | One-Dimensional Quantum Mechanics | 40-1 |
| Chapter 41 | Atomic Physics | 41-1 |
| Chapter 42 | Nuclear Physics | 42-1 |

Dynamics Worksheets
Momentum Worksheets
Energy Worksheets

# Preface

Learning physics, just as learning any skill, requires regular practice of the basic techniques. That is what this *Student Workbook* is all about. The workbook consists of exercises that give you an opportunity to practice the ideas and techniques presented in the textbook and in class. These exercises are intended to be done on a daily basis, right after the topics have been discussed in class and are still fresh in your mind.

You will find that the exercises are nearly all *qualitative* rather than *quantitative*. They ask you to draw pictures, interpret graphs, use ratios, write short explanations, or provide other answers that do not involve significant calculations. The purpose of these exercises is to help you develop the basic thinking tools you'll later need for quantitative problem solving. Successful completion of the workbook exercises will prepare you to tackle the more quantitative end-of-chapter homework problems in the textbook. It is highly recommended that you do the workbook exercises *before* starting the end-of-chapter problems.

You will find that the exercises in this workbook are keyed to specific sections of the textbook in order to let you practice the new ideas introduced in that section. You should keep the text beside you as you work and refer to it often. You will usually find Tactics Boxes, figures, or examples in the textbook that are directly relevant to the exercises. When asked to draw figures or diagrams, you should attempt to draw them so that they look much like the figures and diagrams in the textbook.

Because the exercises go with specific sections of the text, you should answer them on the basis of information presented in *just* that section (and prior sections). You may have learned new ideas in Section 7 of a chapter, but you should not use those ideas when answering questions from Section 4. There will be ample opportunity in the Section 7 exercises to use that information there.

You will need a few "tools" to complete the exercises. Many of the exercises will ask you to *color code* your answers by drawing some items in black, others in red, and yet others in blue. You need to purchase a few colored pencils to do this. The author highly recommends that you work in pencil, rather than ink, so that you can easily erase. Few people produce work so free from errors that they can work in ink! In addition, you'll find that a small, easily-carried six-inch ruler will come in handy for drawings and graphs.

As you work your way through the textbook and this workbook, you will find that physics is a way of *thinking* about how the world works and why things happen as they do. We will be interested primarily in finding relationships and seeking explanations, only secondarily in computing numerical answers. In many ways, the thinking tools developed in this workbook are what the course is all about. If you take the time to do these exercises regularly and to review the answers, in whatever form your instructor provides them, you will be well on your way to success in physics.

**To the instructor:** The exercises in this workbook can be used in many ways. You can have students work on some exercises in class as part of an active-learning strategy. Or you can do the same in recitation sections or laboratories. This approach allows you to discuss the answers immediately, to answer student questions, and to improvise follow-up exercises when needed. Having the students work in small groups (2 to 4 students) is highly recommended.

Alternatively, the exercises can be assigned as homework. The pages are perforated for easy tear-out, and the page breaks are in logical places so that you can assign the sections of a chapter that you would likely cover in one day of class. Exercises should be assigned immediately after presenting the relevant information in class and should be due at the beginning of the next class. Collecting them at the beginning of class, then going over two or three that are likely to cause difficulty, is an effective means of quickly reviewing major concepts from the previous class and launching a new discussion.

If the exercisees are used as homework, it is *essential* for students to receive *prompt* feedback. Ideally this would occur by having the exercises graded, with written comments, and returned at the next class meeting. Posting the answers on a course website also works. Lack of prompt feedback can negate much of the value of these exercises. Placing similar qualitative/ graphical questions on quizzes and exams, and telling students at the beginning of the term that you will do so, encourages students to take the exercises seriously and to check the answers.

The author has been successful with assigning *all* exercises in the workbook as homework, collecting and grading them every day through Chapter 4, then collecting and grading them on about one-third of subsequent days on a random basis. Student feedback from end-of-term questionnaires reveals three prevalent attitudes toward the workbook exercises:

i.   They think it is an unreasonable amount of work.

ii.  They agree that the assignments force them to keep up and not get behind.

iii. They recognize, by the end of the term, that the workbook is a valuable learning tool.

However you choose to use these exercises, they will significantly strengthen your students' conceptual understanding of physics.

Following the workbook exercises are optional Dynamics Worksheets, Momentum Worksheets, and Energy Worksheets for use with end-of-chapter problems in Parts I and II of the textbook. Their use is recommended to help students acquire good problem-solving habits early in the course. End-of-chapter problems marked with the ✎ icon are intended to be done on worksheets.

Answers to all workbook exercises are provided as pdf files on the *Instructor's Supplement* CD-ROM. The author gratefully acknowledges the careful work of answer writers Professor James H. Andrew of Youngstown State University (Chapters 1–15 and 20–24) and Professor Susan Cable of Central Florida Community College (Chapters 16–19 and 25–42).

**Acknowledgments:** Many thanks to Craig Johnson, Simmy Cover, and Jean Lake for handling the production of the *Student Workbook*.

# 1 Concepts of Motion

## 1.1 Motion Diagrams

## 1.2 The Particle Model

**Exercises 1–5:** Draw a motion diagram for each motion described below.
- Use the particle model to represent the object as a particle.
- Six to eight dots are appropriate for most motion diagrams.
- Number the positions in order, as shown in Figure 1.4 in the text.
- Be neat and accurate!

1. A car accelerates forward from a stop sign. It eventually reaches a steady speed of 45 mph.

2. An elevator starts from rest at the 100th floor of the Empire State Building and descends, with no stops, until coming to rest on the ground floor. (Draw this one *vertically* since the motion is vertical.)

3. A skier starts *from rest* at the top of a 30° snow-covered slope and steadily speeds up as she skies to the bottom. (Orient your diagram as seen from the *side*. Label the 30° angle.)

4. The space shuttle orbits the earth in a circular orbit, completing one revolution in 90 minutes.

5. Bob throws a ball at an upward 45° angle from a third-story balcony. The ball lands on the ground below.

**Exercises 6–9:** For each motion diagram, write a short description of the motion of an object that will match the diagram. Your descriptions should name *specific* objects and be phrased similarly to the descriptions of Exercises 1 to 5. Note the axis labels on Exercises 8 and 9.

6.

7.

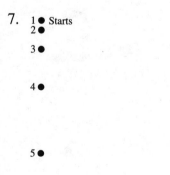

8.

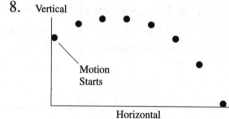

9.

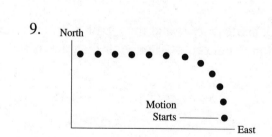

# 1.3  Position and Time

10. The figure below shows the location of an object at three successive instants of time.

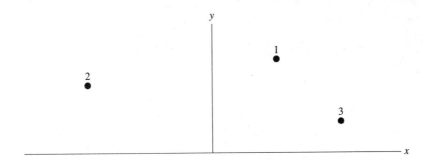

   a. Use a **red** pencil to draw and label on the figure the three position vectors $\vec{r}_1$, $\vec{r}_2$, and $\vec{r}_3$ at times 1, 2, and 3.
   b. Use a **blue** or **green** pencil to draw a possible trajectory from 1 to 2 to 3.
   c. Use a **black** pencil to draw the displacement vector $\Delta\vec{r}$ from the initial to the final position.

11. In Exercise 10, is the object's displacement equal to the distance the object travels? Explain.

12. Redraw your motion diagrams from Exercises 1 to 5 in the space below. Then add and label the displacement vectors $\Delta\vec{r}$ on each diagram.

## 1.4 Velocity

13. The figure below shows the positions of a moving object in three successive frames of film. Draw and label the velocity vector $\vec{v}_1$ for the motion from 1 to 2 and the vector $\vec{v}_2$ for the motion from 2 to 3.

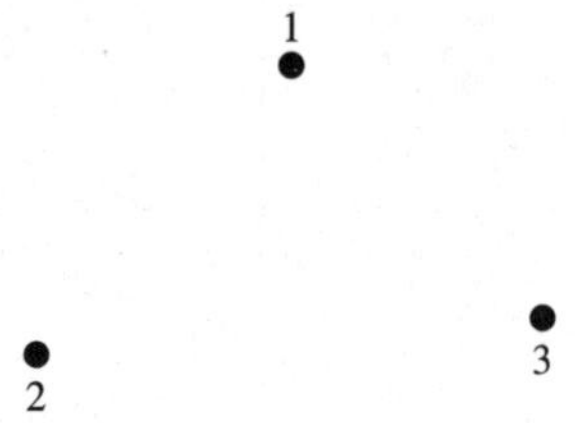

<br>

$\overset{1}{\bullet}$

$\overset{\bullet}{3}$

$\underset{2}{\bullet}$

<br>

**Exercises 14–20:** Draw a motion diagram for each motion described below.
- Use the particle model.
- Show and label the *velocity* vectors.

14. A rocket-powered car on a test track accelerates from rest to a high speed, then coasts at constant speed after running out of fuel. Draw a dotted line across your diagram to indicate the point at which the car runs out of fuel.

15. Galileo drops a ball from the Leaning Tower of Pisa. Consider the ball's motion from the moment it leaves his hand until a microsecond before it hits the ground. Your diagram should be vertical.

16. An elevator starts from rest at the ground floor. It accelerates upward for a short time, then moves with constant speed, and finally brakes to a halt at the tenth floor. Draw dotted lines across your diagram to indicate where the acceleration stops and where the braking begins. You'll need 10 or 12 points to indicate the motion clearly.

17. A bowling ball being returned from the pin area to the bowler starts out rolling at a constant speed. It then goes up a ramp and exits onto a level section at very low speed. You'll need 10 or 12 points to indicate the motion clearly.

18. A track star runs once around a running track at constant speed. The track has straight sides and semicircular ends. Use a bird's-eye view looking down on the track. Use about 20 points for your motion diagram.

19. A car is parked on a hill. The brakes fail, and the car rolls down the hill with an ever-increasing speed. At the bottom of the hill it runs into a thick hedge and gently comes to a halt.

20. Andy is standing on the street. Bob is standing on the second-floor balcony of their apartment, about 30 feet back from the street. Andy throws a baseball to Bob. Consider the ball's motion from the moment it leaves Andy's hand until a microsecond before Bob catches it.

## 1.5 Acceleration

## 1.6 Examples of Motion Diagrams

**Note:** Beginning with this section, and for future motion diagrams, you will "color code" the vectors. Draw velocity vectors **black** and acceleration vectors **red**.

**Exercises 21–26:** The figures below show an object's position in three successive frames of film. The object is moving in the direction $1 \rightarrow 2 \rightarrow 3$. For each diagram:
- Draw and label the initial and final velocity vectors $\vec{v}_0$ and $\vec{v}_1$. Use **black**.
- Use the steps of Tactics Box 1.3 to find the change in velocity $\Delta\vec{v}$.
- Draw and label $\vec{a}$ at the proper location on the motion diagram. Use **red**.
- Determine whether the object is speeding up, slowing down, or moving at a constant speed. Write your answer beside the diagram.

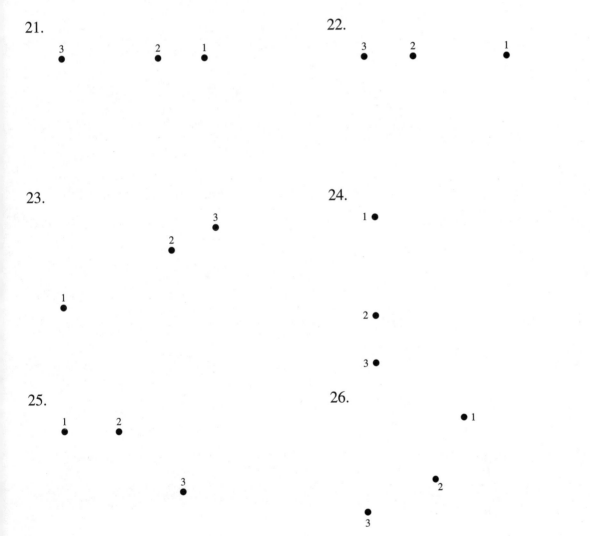

21.

22.

23.

24.

25.

26.

**Exercises 27–34:** Draw a complete motion diagram for each of the following.
- Draw and label the velocity vectors $\vec{v}$. Use **black**.
- Draw and label the acceleration vectors $\vec{a}$. Use **red**.

27. Galileo drops a ball from the Leaning Tower of Pisa. Consider its motion from the moment it leaves his hand until a microsecond before it hits the ground.

28. Trish is driving her car at a steady 30 mph when a small furry creature runs into the road in front of her. She hits the brakes and skids to a stop. Show her motion from 2 seconds before she starts braking until she comes to a complete stop.

29. A ball rolls up a smooth board tilted at a 30° angle. Then it rolls back to its starting position.

30. A bowling ball being returned from the pin area to the bowler rolls at a constant speed, then up a ramp, and finally exits onto a level section at very low speed.

31. A track star runs once around a running track at constant speed. The track has straight sides and semicircular ends. Use a bird's-eye view looking down on the track. Use about 20 dots for this motion diagram.

32. A cannon ball is fired from a Civil War cannon up onto a high cliff. Show the cannon ball's motion from the instant it leaves the cannon until a microsecond before it hits the ground.

33. A plane flying north at 300 mph turns slowly to the west without changing speed, then continues to fly west. Draw the motion diagram from a viewpoint above the plane.

34. Two sprinters, Cynthia and Diane, start side by side. Diane has run only 80 m when Cynthia crosses the finish line of the 100 m dash.

## 1.7  From Words to Symbols

## 1.8  A Problem-Solving Strategy

35. The four motion diagrams below show an initial point 0 and a final point 1. A pictorial representation would define the five symbols: $x_0$, $x_1$, $v_{0x}$, $v_{1x}$, and $a_x$ for horizontal motion and equivalent symbols with $y$ for vertical motion. Determine whether each of these quantities is positive, negative, or zero. Give your answer by writing +, −, or 0 in the table below.

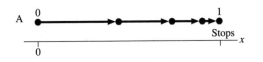

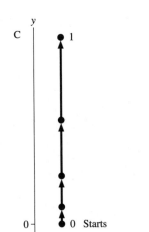

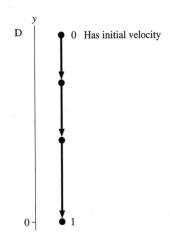

|  | A | B | C | D |
|---|---|---|---|---|
| $x_0$ or $y_0$ |  |  |  |  |
| $x_1$ or $y_1$ |  |  |  |  |
| $v_{0x}$ or $v_{0y}$ |  |  |  |  |
| $v_{1x}$ or $v_{1x}$ |  |  |  |  |
| $a_x$ or $a_y$ |  |  |  |  |

36. The three symbols $x$, $v_x$, and $a_x$ have eight possible combinations of *signs*. For example, one combination is $(x, v_x, a_x) = (+, -, +)$.

    a. List all eight combinations of signs for $x$, $v_x$, $a_x$.

    1. _____    5. _____

    2. _____    6. _____

    3. _____    7. _____

    4. _____    8. _____

b. For each of the eight combinations of signs you identified in part a:
   • Draw a four-dot motion diagram of an object that has these signs for $x$, $v_x$, and $a_x$.
   • Draw the diagram *above* the axis whose number corresponds to part a.
   • Use **black** and **red** for your $\vec{v}$ and $\vec{a}$ vectors. Be sure to label the vectors.

1.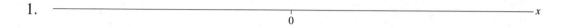
     0                                                                          $x$

2.
     0                                                                          $x$

3.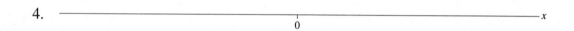
     0                                                                          $x$

4.
     0                                                                          $x$

5.
     0                                                                          $x$

6.
     0                                                                          $x$

7. 
     0                                                                          $x$

8.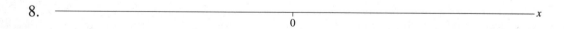
     0                                                                          $x$

# 1.9  Units and Significant Figures

37. Convert the following to SI units. Work across the line and show all steps in the conversion.

   a. $9.12 \, \mu s \times$

   b. $3.42 \, km \times$

   c. $44 \, cm/ms \times$

   d. $80 \, km/hr \times$

   e. $60 \, mph \times$

   f. $8 \, in \times$

   g. $14 \, in^2 \times$

   h. $250 \, cm^3 \times$

   **Note:** Think carefully about g and h. A picture may help.

38. Use Table 1.4 to assess whether or not the following statements are *reasonable*.

   a. Joe is 180 cm tall.

   b. I rode my bike to campus at a speed of 50 m/s.

   c. A skier reaches the bottom of the hill going 25 m/s.

d. I can throw a ball a distance of 2 km.

e. I can throw a ball at a speed of 50 km/hr.

39. Justify the assertion that 1 m/s ≈ 2 mph by *exactly* converting 1 m/s to English units. By what percentage is this rough conversion in error?

40. How many significant figures does each of the following numbers have?

a. 6.21 _____          e. 0.0621 _____          i. 1.0621 _____

b. 62.1 _____          f. 0.620 _____          j. $6.21 \times 10^3$ _____

c. 6210 _____          g. 0.62 _____          k. $6.21 \times 10^{-3}$ _____

d. 6210.0 _____          h. .62 _____          l. $62.1 \times 10^3$ _____

41. Compute the following numbers, applying the significant figure standards adopted for this text.

a. $33.3 \times 25.4 =$ _____          e. $2.345 \times 3.321 =$ _____

b. $33.3 - 25.4 =$ _____          f. $(4.32 \times 1.23) - 5.1 =$ _____

c. $33.3 \div 45.1 =$ _____          g. $33.3^2 =$ _____

d. $33.3 \times 45.1 =$ _____          h. $\sqrt{33.3} =$ _____

# 2 Kinematics: The Mathematics of Motion

## 2.1 Motion in One Dimension

1. Sketch position-versus-time graphs for the following motions. Include a numerical scale on both axes with units that are *reasonable* for this motion. Some numerical information is given in the problem, but for other quantities make reasonable estimates.

   **Note:** A *sketched* graph simply means hand-drawn, rather than carefully measured and laid out with a ruler. But a sketch should still be neat and as accurate as is feasible by hand. It also should include labeled axes and, if appropriate, tick-marks and numerical scales along the axes.

   a. A student walks to the bus stop, waits for the bus, then rides to campus. Assume that all the motion is along a straight street.

   b. A student walks slowly to the bus stop, realizes he forgot his paper that is due, and *quickly* walks home to get it.

   c. The quarterback drops back 10 yards from the line of scrimmage, then throws a pass 20 yards to the tight end, who catches it and sprints 20 yards to the goal. Draw your graph for the *football*. Think carefully about what the slopes of the lines should be.

2. Interpret the following position-versus-time graphs by writing a very short "story" of what is happening. Be creative! Have characters and situations! Simply saying that "a car moves 100 meters to the right" doesn't qualify as a story. Your stories should make *specific reference* to information you obtain from the graphs, such as distances moved or time elapsed.

a. Moving car

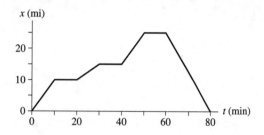

b. Sprinter

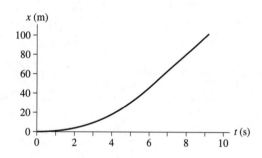

c. Submarine

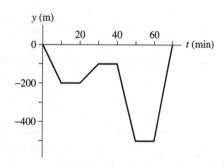

d. Two football players

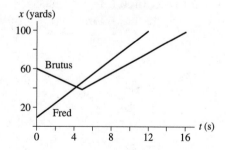

3. Can you give an interpretation to this position-versus-time graph? If so, then do so. If not, why not?

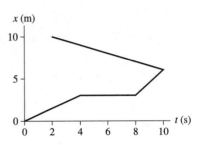

## 2.2 Uniform Motion

4. Sketch position-versus-time graphs for the following motions. Include appropriate numerical scales along both axes. A small amount of computation may be necessary.

a. A parachutist opens her parachute at an altitude of 1500 m. She then descends slowly to earth at a steady speed of 5 m/s. Start your graph as her parachute opens.

b. Trucker Bob starts the day 120 miles west of Denver. He drives east for 3 hours at a steady 60 miles/hour before stopping for his coffee break. Let Denver be located at $x = 0$ mi and assume that the $x$-axis points to the east.

c. Quarterback Bill throws the ball to the right at a speed of 15 m/s. It is intercepted 45 m away by Carlos, who is running to the left at 7.5 m/s. Carlos carries the ball 60 m to score. Let $x = 0$ m be the point where Bill throws the ball. Draw the graph for the *football*.

5. The figure shows a position-versus-time graph for the motion of objects A and B that are moving along the same axis.

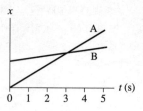

  a. At the instant $t = 1$ s, is the speed of A greater than, less than, or equal to the speed of B? Explain.

  b. Do objects A and B ever have the *same* speed? If so, at what time or times? Explain.

6. Interpret the following position-versus-time graphs by writing a short "story" about what is happening. Your stories should make specific references to the *speeds* of the moving objects, which you can determine from the graphs. Assume that the motion takes place along a horizontal line.

  a.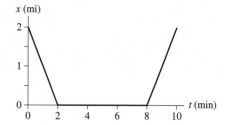

  b.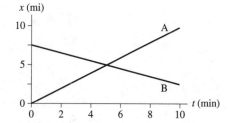

  c.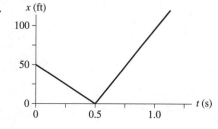

## 2.3 Instantaneous Velocity

7. Draw both a position-versus-time graph *and* a velocity-versus-time graph for an object that is at rest at $x = 1$ m.

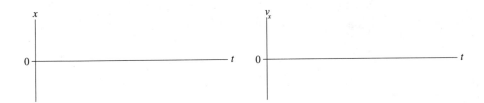

8. The figure shows the position-versus-time graphs for two objects, A and B, that are moving along the same axis.

   a. At the instant $t = 1$ s, is the speed of A greater than, less than, or equal to the speed of B? Explain.

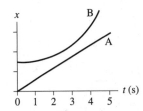

   b. Do objects A and B ever have the *same* speed? If so, at what time or times? Explain.

9. Below are six position-versus-time graphs. For each, draw the corresponding velocity-versus-time graph directly below it. A vertical line drawn through both graphs should connect the velocity $v_s$ at time $t$ with the position $s$ at the *same* time $t$. There are no numbers, but your graphs should correctly indicate the *relative* speeds.

a.

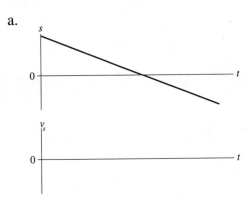

b.

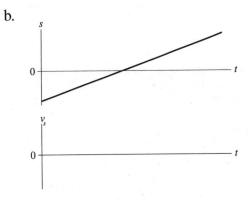

c.

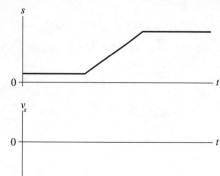

d.

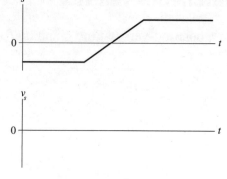

e.

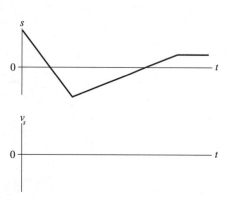

f.

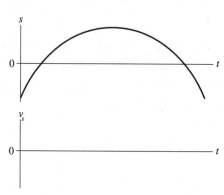

10. The figure shows a position-versus-time graph for a moving object. At which lettered point or points:

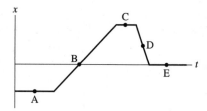

a. Is the object moving the slowest? _____

b. Is the object moving the fastest? _____

c. Is the object at rest? _____

d. Does the object have a constant nonzero velocity? _____

e. Is the object moving to the left? _____

11. The figure shows a position-versus-time graph for a moving object. At which lettered point or points:

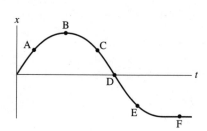

a. Is the object moving the fastest? _____

b. Is the object moving to the left? _____

c. Is the object speeding up? _____

d. Is the object slowing down? _____

e. Is the object turning around? _____

12. For each of the following motions, draw
   • A motion diagram,
   • A position-versus-time graph, and
   • A velocity-versus-time graph.

   a. A car starts from rest, steadily speeds up to 40 mph in 15 s, moves at a constant speed for 30 s, then comes to a halt in 5 s.

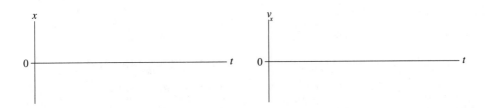

   b. A rock is dropped from a bridge and steadily speeds up as it falls. It is moving at 30 m/s when it hits the ground 3 s later. Think carefully about the signs.

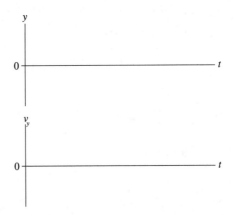

   c. A pitcher winds up and throws a baseball with a speed of 40 m/s. One-half second later the batter hits a line drive with a speed of 60 m/s. The ball is caught 1 s after it is hit. From where you are sitting, the batter is to the right of the pitcher. Draw your motion diagram and graph for the *horizontal* motion of the ball.

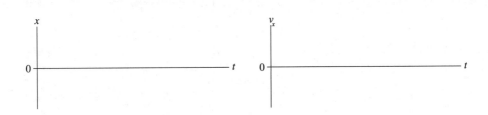

13. The figure shows six frames from the motion diagram of two moving cars, A and B.

    a. Draw both a position-versus-time graph and a velocity-versus-time graph. Show the motion of *both* cars on each graph. Label them A and B.

    b. Do the two cars ever have the same position at one instant of time?

    If so, in which frame number (or numbers)? _____

    Draw a vertical line through your graphs of part a to indicate this instant of time.

    c. Do the two cars ever have the same velocity at one instant of time?

    If so, between which two frames? _____

14. The figure shows six frames from the motion diagram of two moving cars, A and B.

    a. Draw both a position-versus-time graph and a velocity-versus-time graph. Show *both* cars on each graph. Label them A and B.

    b. Do the two cars ever have the same position at one instant of time?

    If so, in which frame number (or numbers)? _____

    Draw a vertical line through your graphs of part a to indicate this instant of time.

    c. Do the two cars ever have the same velocity at one instant of time?

    If so, between which two frames? _____

15. You're driving along the highway at a steady speed of 60 mph when another car decides to pass you. At the moment when the front of his car is exactly even with the front of your car, and you turn your head to smile at him, do the two cars have equal velocities? Explain.

# 2.4 Finding Position from Velocity

16. Below are shown four velocity-versus-time graphs. For each:
   - Draw the corresponding position-versus-time graph.
   - Give a written description of the motion.

   Assume that the motion takes place along a horizontal line and that $x_0 = 0$.

a.

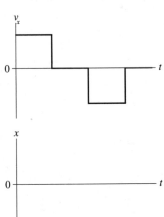

b.

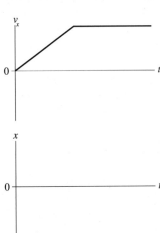

c.

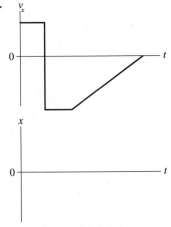

d.

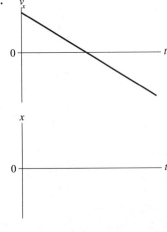

17. The figure shows the velocity-versus-time graph for a moving object whose initial position is $x_0 = 20$ m. Find the object's position graphically, using the geometry of the graph, at the following times.

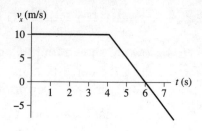

a. At $t = 3$ s.

b. At $t = 5$ s.

c. At $t = 7$ s.

d. You should have found a simple relationship between your answers to parts b and c. Can you explain this? What is the object doing?

## 2.5  Motion with Constant Acceleration

18. A car is traveling north. Can its acceleration vector ever point south? Explain.

19. Give a specific example for each of the following situations. For each, provide:
    - A description, and
    - A motion diagram.
    a.   $a_x = 0$ but $v_x \neq 0$.

    b.   $v_x = 0$ but $a_x \neq 0$.

    c.   $v_x < 0$ and $a_x > 0$.

20. Below are three velocity-versus-time graphs. For each:
    - Draw the corresponding acceleration-versus-time graph.
    - Draw a motion diagram below the graphs.

a.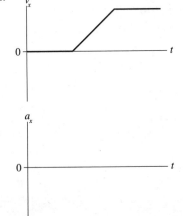

b.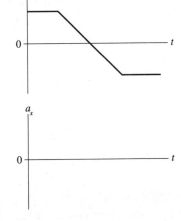

c.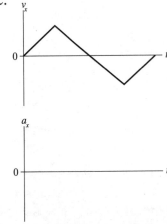

21. Below are three acceleration-versus-time graphs. For each, draw the corresponding velocity-versus-time graph. Assume that $v_{0x} = 0$.

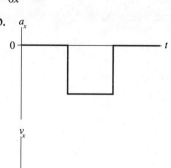
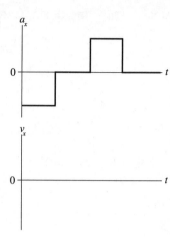

a.   b.   c.

22. The figure below shows nine frames from the motion diagram of two cars. Both cars begin to accelerate, with constant acceleration, in frame 4.

a. Which car has the largest initial velocity? _____ The largest final velocity? _____

b. Which car has the largest acceleration after frame 4? How can you tell?

c. Draw position, velocity, and acceleration graphs, showing the motion of both cars on each graph. (Label them A and B.) This is a total of three graphs with two curves on each.

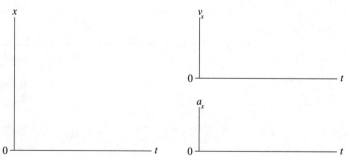

d. Do the cars ever have the same position at one instant of time? If so, in which frame? _____

e. Do the two cars ever have the same velocity at one instant of time?

If so, identify the *two* frames between which this velocity occurs. _____

Identify this instant on your graphs by drawing a vertical line through the graphs.

## 2.6 Free Fall

23. A ball is thrown straight up into the air. At each of the following instants, is the ball's acceleration $g$, $-g$, $0$, $< g$, or $> g$?

   a. Just after leaving your hand?  _____

   b. At the very top (maximum height)?  _____

   c. Just before hitting the ground?  _____

24. A rock is *thrown* (not dropped) straight down from a bridge into the river below.

   a. Immediately *after* being released, is the magnitude of the rock's acceleration greater than $g$, less than $g$, or equal to $g$? Explain.

   b. Immediately before hitting the water, is the magnitude of the rock's acceleration greater than $g$, less than $g$, or equal to $g$? Explain.

25. Alicia throws a red ball straight up into the air, releasing it with velocity $v_0$. As she is throwing it, you happen to pass by in an elevator that is rising with constant velocity $v_0$. At the exact instant Alicia releases her ball, you reach out of the elevator's window (this is a very fancy elevator!) and *gently* release a blue ball. Both balls are the same height above the ground at the moment they are released.

   a. Describe the motion of the two balls as Alicia sees them from the ground. In what ways are the motion of the red ball and the blue ball the same or different?

b. Describe the motion of the two balls as you see them from the moving elevator. In what ways are the motion of the red ball and the blue ball the same or different?

c. Alicia sees a well-defined "top" of the motion where her red ball reaches a maximum height and then starts to fall. Call the time of maximum height $t_1$. As you watch from the elevator, do *you* see anything distinctive or different about the red ball's motion at time $t_1$? If so, what?

d. Does the red ball "stop" at time $t_1$ when Alice sees it at the very top of its trajectory? As part of answering this question, define what you mean by the word "stop."

## 2.7 Motion on an Inclined Plane

26. A ball released from rest on an inclined plane accelerates down the plane at 2 m/s². Complete the table below showing the ball's velocities at the times indicated. Do *not* use a calculator for this; this is a reasoning question, not a calculation problem.

| Time (s) | Velocity (m/s) |
|----------|----------------|
| 0 | 0 |
| 1 | _____ |
| 2 | _____ |
| 3 | _____ |
| 4 | _____ |
| 5 | _____ |

27. A bowling ball rolls along a level surface, then up a 30° slope, and finally exits onto another level surface at a much slower speed.

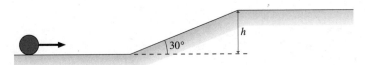

a. Draw position-, velocity-, and acceleration-versus-time graphs for the ball.

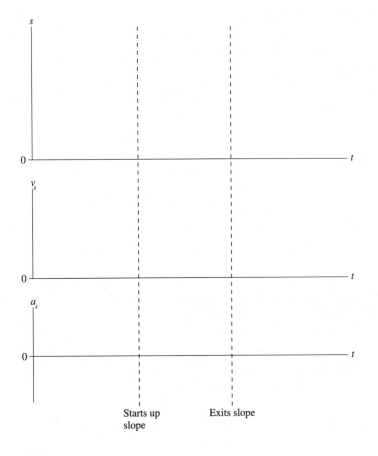

b. Suppose that the ball's initial speed is 5.0 m/s and its final speed is 1.0 m/s. Draw a pictorial representation that you would use to determine the height $h$ of the slope. Establish a coordinate system, define all symbols, list known information, and identify desired unknowns.

**Note:** Don't actually solve the problem. Just draw the complete pictorial representation that you would use as a first step in solving the problem.

## 2.8 Instantaneous Acceleration

28. Below are two acceleration-versus-time curves. For each, draw the corresponding velocity-versus-time curve. Assume that $v_{0x} = 0$.

a.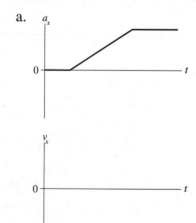

b.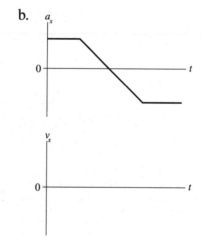

# 3 Vectors and Coordinate Systems

## 3.1 Scalars and Vectors

## 3.2 Properties of Vectors

**Exercises 1–3:** Draw and label the vector sum $\vec{A} + \vec{B}$.

1.

2.

3.

4. Use a figure and the properties of vector addition to show that vector addition is associative. That is, show that

$$(\vec{A} + \vec{B}) + \vec{C} = \vec{A} + (\vec{B} + \vec{C})$$

**Exercises 5–7:** Draw and label the vector difference $\vec{A} - \vec{B}$.

5.

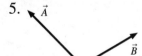

6.

7.

8. Draw and label the vector $2\vec{A}$ and the vector $\frac{1}{2}\vec{A}$.

9. Is it possible to add a scalar to a vector? If so, demonstrate. If not, explain why not.

10. How would you define the *zero vector* $\vec{0}$?

11. Given vectors $\vec{A}$ and $\vec{B}$ below, find the vector $\vec{C} = 2\vec{A} - 3\vec{B}$.

## 3.3 Coordinate Systems and Vector Components

**Exercises 12–14:** Draw and label the *x*- and *y*-component vectors of the vector shown.

12.

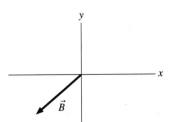

13.

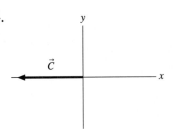

14.

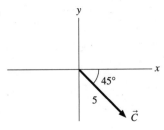

**Exercises 15–17:** Determine the numerical values of the *x*- and *y*-components of each vector.

15.

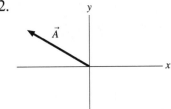

$A_x =$ _____

$A_y =$ _____

16.

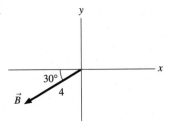

$B_x =$ _____

$B_y =$ _____

17.

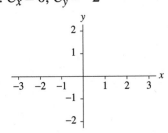

$C_x =$ _____

$C_y =$ _____

**Exercises 18–20:** Draw and label the vector with these components. Then determine the magnitude of the vector.

18. $A_x = 3, A_y = -2$

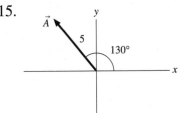

$A =$ _____

19. $B_x = -2, B_y = 2$

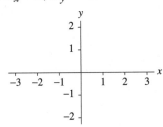

$B =$ _____

20. $C_x = 0, C_y = -2$

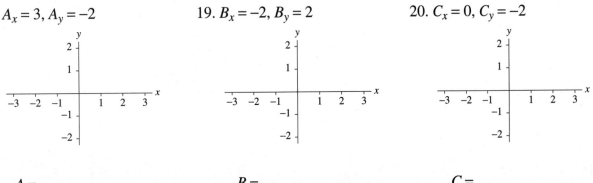

$C =$ _____

21. Can a vector have a component equal to zero and still have nonzero magnitude? Explain.

22. Can a vector have zero magnitude if one of its components is nonzero? Explain.

23. How would you define the *zero vector* $\vec{0}$ by using the idea of components?

24. Suppose two vectors have unequal magnitudes. Can their sum be zero? Explain.

# 3.4 Vector Algebra

**Exercises 25–27:** Draw and label the vectors on the axes.

25. $\vec{A} = -\hat{i} + 2\hat{j}$

26. $\vec{B} = -2\hat{j}$

27. $\vec{C} = 3\hat{i} - 2\hat{j}$

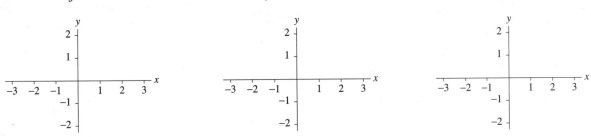

**Exercises 28–30:** Write the vector in component form (e.g., $3\hat{i} + 2\hat{j}$).

28.

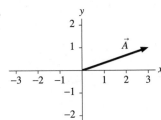

$\vec{A} = $ _____

29.

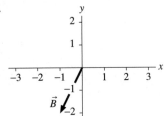

$\vec{B} = $ _____

30.

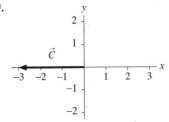

$\vec{C} = $ _____

31. What is the vector sum $\vec{D} = \vec{A} + \vec{B} + \vec{C}$ of the three vectors defined in Exercises 28–30? Write your answer in *component* form.

**Exercises 32–34:** For each vector:
- Draw the vector on the axes provided.
- Draw and label an angle $\theta$ to describe the direction of the vector.
- Find the magnitude and the angle of the vector.

32. $\vec{A} = 2\hat{i} + 2\hat{j}$

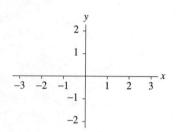

$A = \underline{\hspace{2cm}}$

$\theta = \underline{\hspace{2cm}}$

33. $\vec{B} = -2\hat{i} + 2\hat{j}$

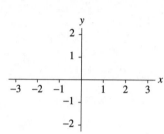

$B = \underline{\hspace{2cm}}$

$\theta = \underline{\hspace{2cm}}$

34. $\vec{C} = 3\hat{i} - \hat{j}$

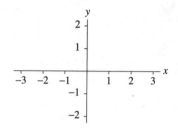

$C = \underline{\hspace{2cm}}$

$\theta = \underline{\hspace{2cm}}$

**Exercises 35–37:** Define vector $\vec{A} = (5, 30°$ above the horizontal$)$. Determine the components $A_x$ and $A_y$ in the three coordinate systems shown below. Show your work below the figure.

35.

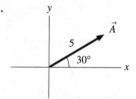

$A_x = \underline{\hspace{2cm}}$

$A_y = \underline{\hspace{2cm}}$

36.

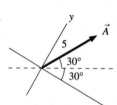

$A_x = \underline{\hspace{2cm}}$

$A_y = \underline{\hspace{2cm}}$

37.

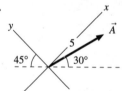

$A_x = \underline{\hspace{2cm}}$

$A_y = \underline{\hspace{2cm}}$

# 4 Force and Motion

## 4.1 Force

1. Two or more forces are shown on the objects below. Draw and label the net force $\vec{F}_{net}$.

 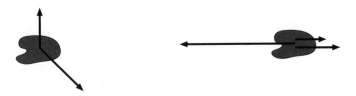

2. Two or more forces are shown on the objects below. Draw and label the net force $\vec{F}_{net}$.

## 4.2 A Short Catalog of Forces

## 4.3 Identifying Forces

**Exercises 3–8:** Follow the six-step procedure of Tactics Box 4.2 to identify and name all the forces acting on the object.

3. An elevator suspended by a cable is descending at constant velocity.

4. A car on a *very* slippery icy road is sliding headfirst into a snowbank, where it gently comes to rest with no one injured. (Question: What does "*very* slippery" imply?)

5. A compressed spring is pushing a block across a rough horizontal table.

6. A brick is falling from the roof of a three-story building.

7. Blocks A and B are connected by a string passing over a pulley. Block B is falling and dragging block A across a frictionless table. Let block A be "the system" for analysis.

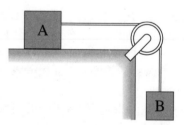

8. A rocket is launched at a 30° angle. Air resistance is not negligible.

# 4.4  What Do Forces Do? A Virtual Experiment

9. The figure shows an acceleration-versus-force graph for an object of mass $m$. Data have been plotted as individual points, and a line has been drawn through the points.

   Draw and label, directly on the figure, the acceleration-versus-force graphs for objects of mass

   a. $2m$                    b. $0.5m$

   Use triangles ▲ to show four points for the object of mass $2m$, then draw a line through the points. Use squares ■ for the object of mass $0.5m$.

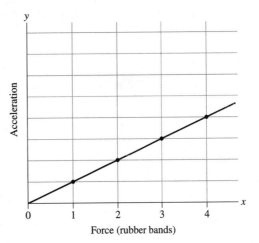

Force (rubber bands)

10. A constant force applied to object A causes A to accelerate at 5 m/s². The same force applied to object B causes an acceleration of 3 m/s². Applied to object C, it causes an acceleration of 8 m/s².

    a. Which object has the largest mass? _____

    b. Which object has the smallest mass? _____

    c. What is the ratio of mass A to mass B? $(m_A/m_B) =$ _____

11. A constant force applied to an object causes the object to accelerate at 10 m/s². What will the acceleration of this object be if

    a. The force is doubled? _____     b. The mass is doubled? _____

    c. The force is doubled *and* the mass is doubled? _____

    d. The force is doubled *and* the mass is halved? _____

12. A constant force applied to an object causes the object to accelerate at 8 m/s². What will the acceleration of this object be if

    a. The force is halved? _____     b. The mass is halved? _____

    c. The force is halved *and* the mass is halved? _____

    d. The force is halved *and* the mass is doubled? _____

# 4.5  Newton's Second Law

13. Forces are shown on two objects. For each:
    a. Draw and label the net force vector. Do this right on the figure.
    b. Below the figure, draw and label the object's acceleration vector.

14. Forces are shown on two objects. For each:
    a. Draw and label the net force vector. Do this right on the figure.
    b. Below the figure, draw and label the object's acceleration vector.

15. In the figures below, one force is missing. Use the given direction of acceleration to determine the missing force and draw it on the object. Do all work directly on the figure.

16. Below are two motion diagrams for a particle. Draw and label the net force vector at point 3.

17. Below are two motion diagrams for a particle. Draw and label the net force vector at point 3.

## 4.6 Newton's First Law

18. If an object is at rest, can you conclude that there are no forces acting on it? Explain.

19. If a force is exerted on an object, is it possible for that object to be moving with constant velocity? Explain.

20. A hollow tube forms three-quarters of a circle. It is lying flat on a table. A ball is shot through the tube at high speed. As the ball emerges from the other end, does it follow path A, path B, or path C? Explain your reasoning.

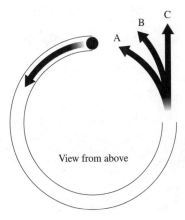

View from above

21. Which, if either, of the objects shown below is in equilibrium? Explain your reasoning.

22. Two forces are shown on the objects below. Add a third force $\vec{F}_3$ that will cause the object to be in equilibrium.

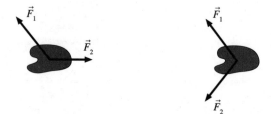

23. Are the following inertial reference frames? Answer Yes or No.

a. A car driving at steady speed on a straight and level road.  _____

b. A car driving at steady speed up a 10° incline.  _____

c. A car speeding up after leaving a stop sign.  _____

d. A car driving at steady speed around a curve.  _____

e. A hot air balloon rising straight up at steady speed.  _____

f. A skydiver just after leaping out of a plane.  _____

g. The space shuttle orbiting the earth.  _____

# 4.7 Free-Body Diagrams

**Exercises 24–29:**
- Draw a picture and identify the forces, then
- Draw a complete free-body diagram for the object, following each of the steps given in Tactics Box 4.3. Be sure to think carefully about the direction of $\vec{F}_{net}$.

**Note:** Draw individual force vectors with a **black** or **blue** pencil or pen. Draw the *net* force vector $\vec{F}_{net}$ with a **red** pencil or pen.

24. A heavy crate is being lowered straight down at a constant speed by a steel cable.

25. A boy is pushing a box across the floor at a steadily increasing speed. Let the box be "the system" for analysis.

26. A bicycle is speeding up down a hill. Friction is negligible, but air resistance is not.

27. You've slammed on your car brakes while going down a hill. You're skidding to a halt.

28. You are going to toss a rock *straight up* into the air by placing it on the palm of your hand (you're not gripping it), then pushing your hand up very rapidly. You may want to toss an object into the air this way to help you think about the situation. The rock is "the system" of interest.

    a. As you hold the rock at rest on your palm, before moving your hand.

    b. As your hand is moving up but before the rock leaves your hand.

    c. One-tenth of a second after the rock leaves your hand.

    d. After the rock has reached its highest point and is now falling straight down.

29. Block B has just been released and is beginning to fall. Consider block A to be "the system."

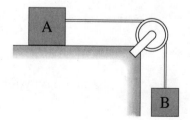

# 5 Dynamics I: Motion Along a Line

## 5.1 Equilibrium

1. The vectors below show five forces that can be applied individually or in combinations to an object. Which forces or combinations of forces will cause the object to be in equilibrium?

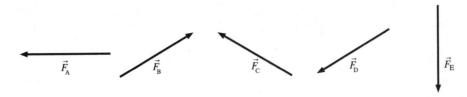

2. Are the objects described here in static equilibrium, dynamic equilibrium, or not in equilibrium at all? Answer Static, Dynamic, or Not.

   a. A girder is lifted at constant speed by a crane. _____

   b. A girder is lowered into place by a crane. It is slowing down. _____

   c. You're straining to hold a 200 pound barbell over your head. _____

   d. A jet plane has reached its cruising speed and altitude. _____

   e. A rock is falling into the Grand Canyon. _____

   f. A box in the back of a truck doesn't slide as the truck stops. _____

3. The free-body diagrams show a force or forces acting on an object. Draw and label one more force (one that is appropriate to the situation) that will cause the object to be in equilibrium.

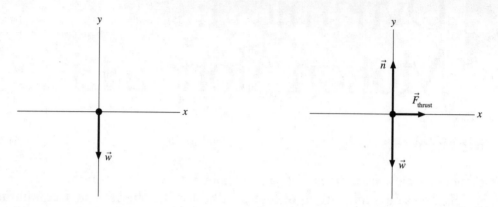

4. The free-body diagrams show a force or forces acting on an object. Draw and label one more force (one that is appropriate to the situation) that will cause the object to be in equilibrium.

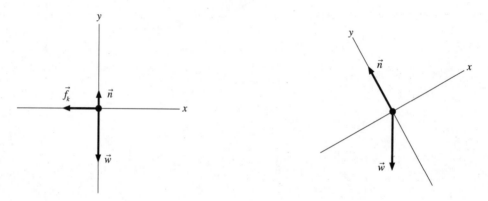

5. Write two or three sentences explaining why you agree or disagree with the statement: "Forces cause an object to move."

6. If you know all of the forces acting on a moving object, can you tell in which direction the object is moving? If the answer is Yes, explain how. If the answer is No, give an example.

## 5.2  Using Newton's Second Law

7. a. An elevator travels *upward* at a constant speed. The elevator hangs by a single cable. Friction and air resistance are negligible. Is the tension in the cable greater than, less than, or equal to the weight of the elevator? Explain. Your explanation should include both a free-body diagram and reference to appropriate physical principles.

   b. The elevator travels *downward* and is slowing down. Is the tension in the cable greater than, less than, or equal to the weight of the elevator? Explain.

**Exercises 8–9:** The figures show free-body diagrams for an object of mass $m$. Write the $x$- and $y$-components of Newton's second law. Write your equations in terms of the *magnitudes* of the forces $F_1, F_2, \ldots$ and any *angles* defined in the diagram. One equation is shown to illustrate the procedure.

8.

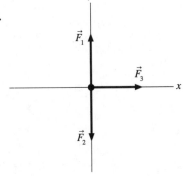

$ma_x =$

$ma_y = F_1 - F_2$

$ma_x =$

$ma_y =$

9.

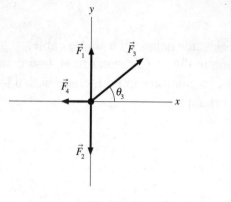   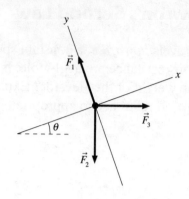

$$ma_x = F_3 \cos \theta_3 - F_4$$

$$ma_y =$$

$$ma_x =$$

$$ma_y =$$

**Exercises 10–12:** Two or more forces, shown on a free-body diagram, are exerted on a 2 kg object. The units of the grid are newtons. For each:

- Draw a vector arrow *on the grid,* starting at the origin, to show the net force $\vec{F}_{net}$.
- In the space to the right, determine the numerical values of the components $a_x$ and $a_y$.

10.

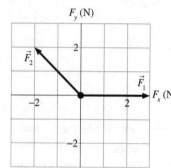

$$a_x =$$

$$a_y =$$

11.

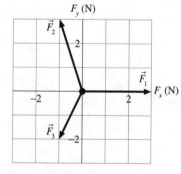

$$a_x =$$

$$a_y =$$

12.

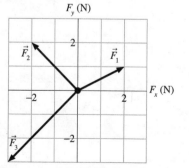

$$a_x =$$

$$a_y =$$

**Exercises 13–15:** Three forces $\vec{F}_1$, $\vec{F}_2$, and $\vec{F}_3$ cause a 1 kg object to accelerate with the acceleration given. Two of the forces are shown on the free-body diagrams below, but the third is missing. For each, draw and label *on the grid* the missing third force vector.

13. $\vec{a} = 2\hat{i} \text{ m/s}^2$

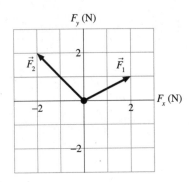

14. $\vec{a} = -3\hat{j}\,\text{m/s}^2$

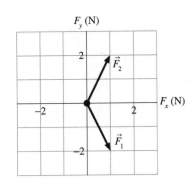

15. The object moves with constant velocity.

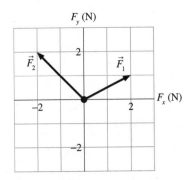

16. Three arrows are shot horizontally. They have left the bow and are traveling parallel to the ground. Air resistance is negligible. Rank in order, from largest to smallest, the magnitudes of the *horizontal* forces $F_1$, $F_2$, and $F_3$ acting on the arrows. Some may be equal. Give your answer in the form A > B = C > D.

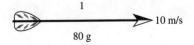
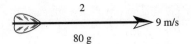
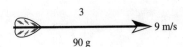

Order:

Explanation:

## 5.3  Mass and Weight

17. Decide whether each of the following is True or False. Give a reason!

    a. The mass of an object depends on its location.

    b. The weight of an object depends on its location.

    c. Mass and weight describe the same thing in different units.

18. An astronaut takes his bathroom scales to the moon and then stands on them. Is the reading of the scales his true weight? Explain.

19. Four balls are thrown straight up. They have the same size, but different mass. Air resistance is negligible. Rank in order, from largest to smallest, the magnitude of the net force acting on each ball. Some may be equal. Give your answer in the form A > B = C > D.

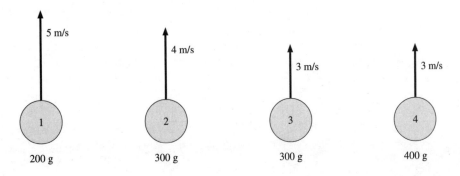

Order:

Explanation:

20. The terms "vertical" and "horizontal" are frequently used in physics. Give *operational definitions* for these two terms. An operational definition defines a term by how it is measured or determined. Your definition should apply equally well in a laboratory or on a steep mountainside.

21. Suppose you have a jet-powered flying platform that can move straight up and down. For each of the following cases, is your apparent weight equal to, greater than, or less than your true weight? Explain.

    a. You are ascending and speeding up.

    b. You are descending and speeding up.

    c. You are ascending at a constant speed.

    d. You are ascending and slowing down.

    e. You are descending and slowing down.

22. Suppose you attempt to pour out 100 g of salt, using a pan balance for measurement, while in an elevator that is accelerating upward. Will the quantity of salt be too much, too little, or the correct amount? Explain.

23. A box with a 75 kg passenger inside is launched straight up into the air by a giant rubber band. After the box has left the rubber band but is still moving *upward:*

    a. What is the passenger's true weight?

    b. What is the passenger's apparent weight?

24. An astronaut orbiting the earth is handed two balls that are identical in outward appearance. However, one is hollow while the other is filled with lead. How might the astronaut determine which is which? Cutting them open is not allowed.

25. Suppose you stand on a spring scale in six identical elevators. Each elevator moves as shown below. Let the reading of the scale in elevator $n$ be $S_n$. Rank in order, from largest to smallest, the six scale readings $S_1$ to $S_6$. Some may be equal. Give your answer in the form $A > B = C > D$.

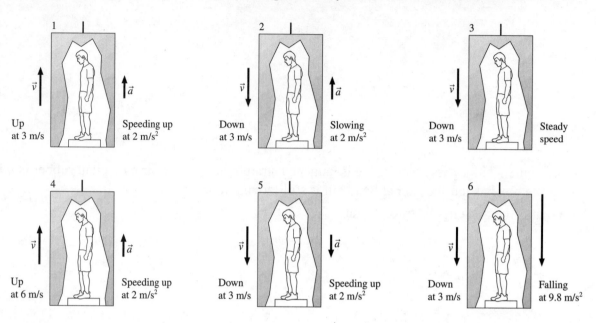

Order:

Explanation:

# 5.4 Friction

26. A block pushed along the floor with velocity $\vec{v}_0$ slides a distance $d$ after the pushing force is removed.

   a. If the mass of the block is doubled but the initial velocity is not changed, what is the distance the block slides before stopping? Explain.

   b. If the initial velocity of the block is doubled to $2\vec{v}_0$ but the mass is not changed, what is the distance the block slides before stopping? Explain.

27. Suppose you press a book against the wall with your hand. The book is not moving.

   a. Identify the forces on the book and draw a free-body diagram.

   b. Now suppose you decrease your push, but not enough for the book to slip. What happens to each of the following forces? Do they increase in magnitude, decrease, or not change?

   $\vec{F}_{push}$  _____

   $\vec{w}$  _____

   $\vec{n}$  _____

   $\vec{f}_s$  _____

   $f_{s\ max}$  _____

28. Consider a box in the back of a pickup truck.

   a. If the truck accelerates slowly, the box moves with the truck without slipping. What force or forces act on the box to accelerate it? In what direction do those forces point?

   b. Draw a free-body diagram of the box.

   c. What happens to the box if the truck accelerates too rapidly? *Explain* why this happens, basing your explanation on physical models and the principles described in this chapter.

# 5.5 Drag

29. Three objects move through the air as shown. Rank in order, from largest to smallest, the three drag forces $D_1$, $D_2$, and $D_3$. Some may be equal. Give your answer in the form A > B = C > D.

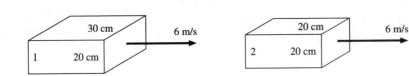

Order:

Explanation:

30. Five balls move through the air as shown. All five have the same size and shape. Rank in order, from largest to smallest, the size of their accelerations $a_1$ to $a_5$. Some may be equal. Give your answer in the form A > B = C > D.

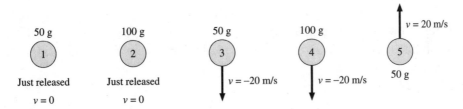

Order:

Explanation:

31. A 1 kg wood ball and a 10 kg lead ball have identical shapes and sizes. They are dropped simultaneously from a tall tower.

   a. To begin, assume that air resistance is negligible. As the balls fall, are the forces on them equal in magnitude or different? If different, which has the larger force? Explain.

**5-14** CHAPTER 5 · Dynamics I: Motion Along a Line

b. Are their accelerations equal or different? If different, which has the larger acceleration? Explain.

c. Which ball hits the ground first? Or do they hit simultaneously? Explain.

d. If air resistance is present, each ball will experience the *same* drag force because both have the same shape. Draw free-body diagrams for the two balls as they fall in the presence of air resistance. Make sure that your vectors all have the correct *relative* lengths.

e. When air resistance is included, are the accelerations of the balls equal or different? If not, which has the larger acceleration? Explain, using your free-body diagrams and Newton's laws.

f. Which ball now hits the ground first? Or do they hit simultaneously? Explain.

## 5.6 More Examples of the Second Law

No exercises.

# 6 Dynamics II: Motion in a Plane

## 6.1 Kinematics in Two Dimensions

1. Complete the motion diagram for this trajectory, showing velocity and acceleration vectors.

2. A particle moving along a trajectory in the *xy*-plane has the *x*-versus-*t* graph and the *y*-versus-*t* graph shown below.

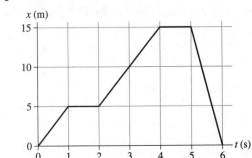 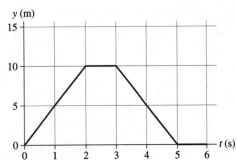

   a. Use the grid below to draw a *y*-versus-*x* graph of the trajectory.

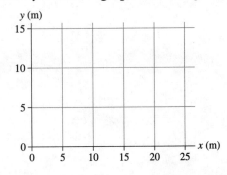

   b. Draw the particle's velocity vector at *t* = 3.5 s on your graph.

3. The trajectory of a particle is shown below. The particle's position is indicated with dots at 1 second intervals. The particle moves between each pair of dots at constant speed.

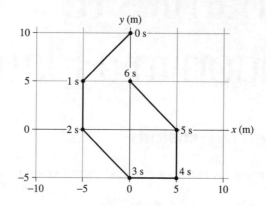

a. Draw *x*-versus-*t* and *y*-versus-*t* graphs for the particle.

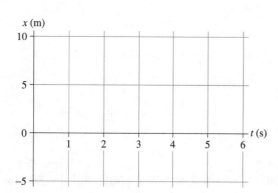 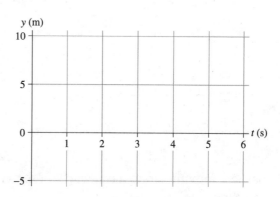

b. Is the particle's speed between *t* = 5 s and *t* = 6 s greater than, less than, or equal to its speed between *t* = 1 s and *t* = 2 s? Explain.

4. The figure shows a ramp and a ball that rolls along the ramp. Draw vector arrows on the figure to show the ball's acceleration at each of the lettered points A to E (or write $\vec{a} = \vec{0}$, if appropriate).

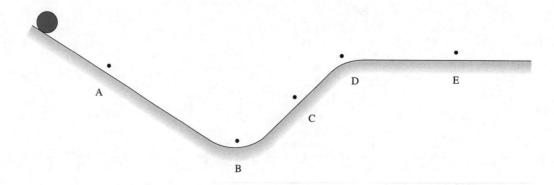

## 6.2 Dynamics in Two Dimensions

5. An ice hockey puck is pushed across frictionless ice in the direction shown. The puck receives a sharp, very short-duration kick toward the right as it crosses line 2. It receives a second kick, of equal strength and duration but toward the left, as it crosses line 3. Sketch the puck's trajectory from line 1 until it crosses line 4.

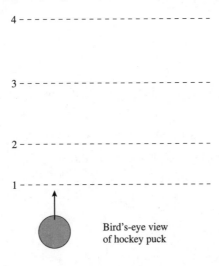

Bird's-eye view of hockey puck

6. A rocket motor is taped to an ice hockey puck, oriented so that the thrust is to the left. The puck is given a push across frictionless ice in the direction shown. The rocket will be turned on by remote control as the puck crosses line 2, then turned off as it crosses line 3. Sketch the puck's trajectory from line 1 until it crosses line 4.

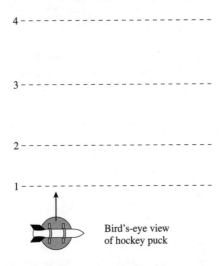

Bird's-eye view of hockey puck

7. An ice hockey puck is sliding from west to east across frictionless ice. When the puck reaches the point marked by the dot, you're going to give it *one* sharp blow with a hammer. After hitting it, you want the puck to move from north to south at a speed similar to its initial west-to-east speed. Draw a force vector with its tail on the dot to show the direction in which you will aim your hammer blow.

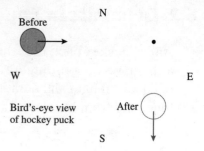

8. Tarzan swings through the jungle by hanging from a vine.

   a. Draw a motion diagram of Tarzan, as you learned in Chapter 1. Use it to find the direction of Tarzan's acceleration vector $\vec{a}$:

      i.   Immediately after stepping off the branch, and
      ii.  At the lowest point in his swing.

   b. At the lowest point in the swing, is the tension $T$ in the vine greater than, less than, or equal to Tarzan's weight? Explain, basing your explanation on Newton's laws.

## 6.3 Projectile Motion

9. A projectile is launched over horizontal ground at an angle between 0° and 90°.

   a. Is there any point on the trajectory where $\vec{v}$ and $\vec{a}$ are parallel to each other? If so, where?

   b. Is there any point where $\vec{v}$ and $\vec{a}$ are perpendicular to each other? If so, where?

   c. Which of the following remain constant throughout the entire trajectory: $r$, $x$, $y$, $v$, $v_x$, $v_y$, $a_x$, $a_y$?

10. The figure shows a ball that rolls down a quarter-circle ramp, then off a cliff. Sketch the ball's trajectory from the instant it is released until it hits the ground.

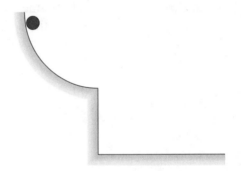

11. a. A cart that is rolling at constant velocity fires a ball straight up. When the ball comes back down, will it land in front of the launching tube, behind the launching tube, or directly in the tube? Explain.

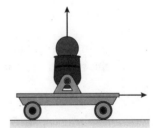

   b. Will your answer change if the cart is accelerating in the forward direction? If so, how?

12. A rock is thrown from a bridge at an angle 30° below horizontal.

    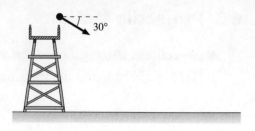

    a. Sketch the rock's trajectory on the figure.

    b. Immediately after the rock is released, is the magnitude of its acceleration greater than, less than, or equal to $g$? Explain.

    c. At the instant of impact, is the rock's speed greater than, less than, or equal to the speed with which it was thrown? Explain.

13. Four balls are simultaneously launched with the same speed from the same height $h$ above the ground. At the same instant, ball 5 is released from rest at the same height. Rank in order, from shortest to longest, the amount of time it takes each of these balls to hit the ground. Ignore air resistance. (Some may be simultaneous.)

    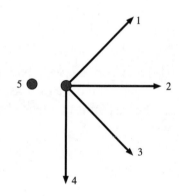

    Order:

    Explanation:

14. Rank in order, from shortest to longest, the amount of time it takes each of these projectiles to hit the ground. Ignore air resistance. (Some may be simultaneous.)

    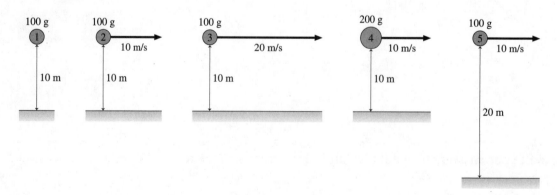

    Order:

    Explanation:

## 6.4 Relative Motion

15. Anita is running to the right at 5 m/s. Balls 1 and 2 are thrown toward her at 10 m/s by friends standing on the ground. According to Anita, which ball is moving faster? Or are both speeds the same? Explain.

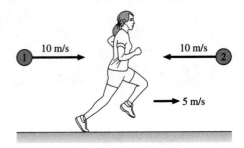

16. Anita is running to the right at 5 m/s. Balls 1 and 2 are thrown toward her by friends standing on the ground. According to Anita, both balls are approaching her at 10 m/s. Which ball was thrown at a faster speed? Or were they thrown with the same speed? Explain.

17. Ryan, Samantha, and Tomas are driving their convertibles. At the same instant, they each see a jet plane with an instantaneous velocity of 200 m/s and an acceleration of 5 m/s².

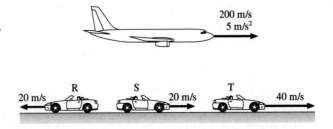

a. Rank in order, from largest to smallest, the jet's *speed* $v_R$, $v_S$, and $v_T$ according to Ryan, Samantha, and Tomas. Explain.

b. Rank in order, from largest to smallest, the jet's *acceleration* $a_R$, $a_S$, and $a_T$ according to Ryan, Samantha, and Tomas. Explain.

18. An electromagnet on the ceiling of an airplane holds a steel ball. When a button is pushed, the magnet releases the ball. The experiment is first done while the plane is parked on the ground, and the point where the ball hits the floor is marked with an X. Then the experiment is repeated while the plane is flying level at a steady 500 mph. Does the ball land slightly in front of the X (toward the nose of the plane), on the X, or slightly behind the X (toward the tail of the plane)? Explain.

19. Zack is driving past his house. He wants to toss his physics book out the window and have it land in his driveway. If he lets go of the book exactly as he passes the end of the driveway, should he direct his throw outward and toward the front of the car (throw 1), straight outward (throw 2), or outward and toward the back of the car (throw 3)? Explain. (Ignore air resistance.)

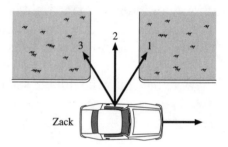

20. Yvette and Zack are driving down the freeway side by side with their windows rolled down. Zack wants to toss his physics book out the window and have it land in Yvette's front seat. Should he direct his throw outward and toward the front of the car (throw 1), straight outward (throw 2), or outward and toward the back of the car (throw 3)? Explain. (Ignore air resistance.)

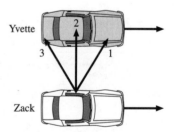

# 7 Dynamics III: Motion in a Circle

## 7.1 Uniform Circular Motion

## 7.2 Velocity and Acceleration in Uniform Circular Motion

1. a. The crankshaft in your car rotates at 3000 rpm. What is the frequency in revolutions per second?

   b. A record turntable rotates at 33.3 rpm. What is the period in seconds?

2. The figure shows three points on a steadily rotating wheel.
   a. Draw the velocity vectors at each of the three points.
   b. Rank in order, from largest to smallest, the angular velocities $\omega_1$, $\omega_2$, and $\omega_3$ of these points.

   Order:

   Explanation:

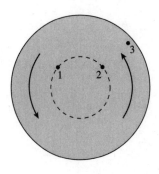

   c. Rank in order, from largest to smallest, the speeds $v_1$, $v_2$, and $v_3$ of these points.

   Order:

   Explanation:

3. Below are two angular position-versus-time graphs. For each, draw the corresponding angular velocity-versus-time graph directly below it.

a.

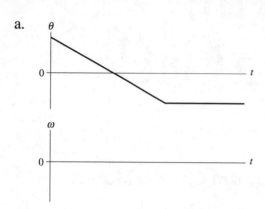

b.

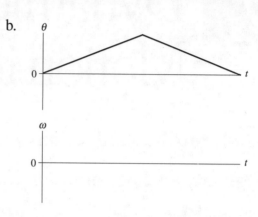

4. Below are two angular velocity-versus-time graphs. For each, draw the corresponding angular position-versus-time graph directly below it. Assume $\theta_0 = 0$ rad.

a.

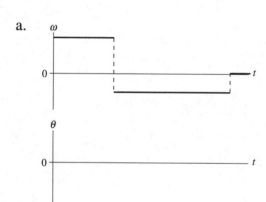

b.

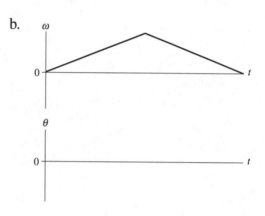

5. A particle in circular motion rotates clockwise at 4 rad/s for 2 s, then counterclockwise at 2 rad/s for 4 s. The time required to change direction is negligible. Graph the angular velocity and the angular position, assuming $\theta_0 = 0$ rad.

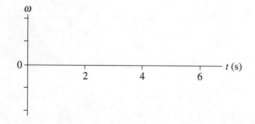
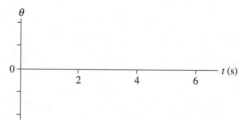

6. A particle rotates in a circle with $a_r = 8$ m/s$^2$. What is $a_r$ if

a. The radius is doubled without changing the angular velocity? _____

b. The radius is doubled without changing the particle's speed? _____

c. The angular velocity is doubled without changing the particle's radius? _____

# 7.3 Dynamics of Uniform Circular Motion

7. The figure shows a *top view* of a plastic tube that is fixed on a horizontal table top. A marble is shot into the tube at A. Sketch the marble's trajectory after it leaves the tube at B.

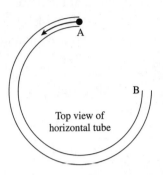

Top view of horizontal tube

8. A ball swings in a *vertical* circle on a string. During one revolution, a very sharp knife is used to cut the string at the instant when the ball is at its lowest point. Sketch the subsequent trajectory of the ball until it hits the ground.

9. The figures are a bird's-eye view of particles moving in horizontal circles on a table top. All are moving at the same speed. Rank in order, from largest to smallest, the tensions $T_1$ to $T_4$.

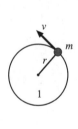
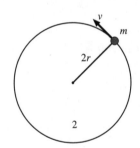
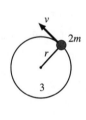
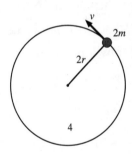

Order:

Explanation:

10. A ball on a string moves in a vertical circle. When the ball is at its lowest point, is the tension in the string greater than, less than, or equal to the ball's weight? Explain. (You may want to include a free-body diagram as part of your explanation.)

11. A marble rolls around the inside of a cone. Draw a free-body diagram of the marble when it is on the left side of the cone and a free-body diagram of the marble when it is on the right side of the cone.

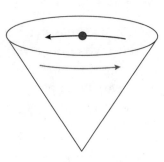

        On left side                On right side

12. A jet airplane is flying on a level course at constant velocity.

    a. What is the *net* force on the plane?   _____

    b. Draw a picture and identify all of the forces acting on the plane.

    c. Airplanes bank when they turn. Explain why, in terms of forces and physical laws. Hint: What would a free-body diagram look like to an observer *behind* the plane?

## 7.4 Circular Orbits

13. The earth has seasons because the axis of the earth's rotation is tilted 23° away from a line perpendicular to the plane of the earth's orbit. You can see this in the figure, which shows the edge of the earth's orbit around the sun. For both positions of the earth, draw a force vector to show the net force acting on the earth or, if appropriate, write $\vec{F} = \vec{0}$.

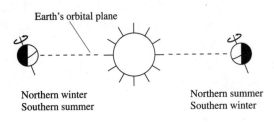

Earth's orbital plane

Northern winter           Northern summer
Southern summer          Southern winter

14. A small projectile is launched parallel to the ground at height $h = 1$ m with sufficient speed to orbit a completely smooth, airless planet. A bug rides in a small hole inside the projectile. Is the bug weightless? Explain.

## 7.5 Fictitious Forces and Apparent Weight

15. A stunt plane does a series of vertical loop-the-loops. At what point in the circle does the pilot feel the heaviest? Explain. Include a free-body diagram with your explanation.

16. A roller-coaster car goes around the inside of a loop-the-loop. Check the statement that is true when the car is at the highest point and at the lowest point in the loop.

|  | Highest | Lowest |
|---|---|---|
| The apparent weight $w_{app}$ is always less than $w$ | _____ | _____ |
| The apparent weight $w_{app}$ is always equal to $w$ | _____ | _____ |
| The apparent weight $w_{app}$ is always greater than $w$ | _____ | _____ |
| $w_{app}$ could be less than, equal to, or greater than $w$ | _____ | _____ |

17. You can swing a ball on a string in a *vertical* circle if you swing it fast enough.

   a. Draw two free-body diagrams of the ball at the top of the circle. On the left, show the ball when it is going around the circle very fast. On the right, show the ball as it goes around the circle more slowly.

   Very fast                                              Slower

   b. As you continue slowing the swing, there comes a frequency at which the string goes slack and the ball doesn't make it to the top of the circle. What condition must be satisfied for the ball to be able to complete the full circle?

   c. Suppose the ball has the smallest possible frequency that allows it to go all the way around the circle. What is the tension in the string when the ball is at the highest point? Explain.

18. It's been proposed that future space stations create "artificial gravity" by rotating around an axis. (The space station would have to be much larger than the present space station for this to be feasible.)

   a. How would this work? Explain.

   b. Would the artificial gravity be equally effective throughout the space station? If not, where in the space station would the residents want to live and work?

# 7.6  Nonuniform Circular Motion

19. For each, figure determine the signs (+ or −) of $\omega$ and $a_t$.

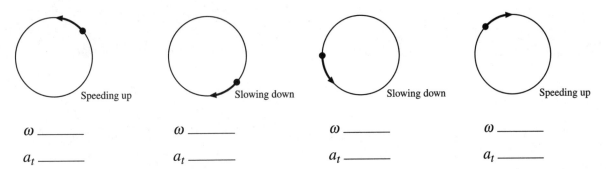

$\omega$ _____

$a_t$ _____

$\omega$ _____

$a_t$ _____

$\omega$ _____

$a_t$ _____

$\omega$ _____

$a_t$ _____

20. The figures below show the radial acceleration vector $\vec{a}_r$ at four sequential points on the trajectory of a particle moving in a counterclockwise circle.

   a. For each, draw the tangential acceleration vector $\vec{a}_t$ at points 2 and 3 or, if appropriate, write $\vec{a}_t = \vec{0}$.

   b. Determine whether $a_t$ is positive (+), negative (−), or zero (0).

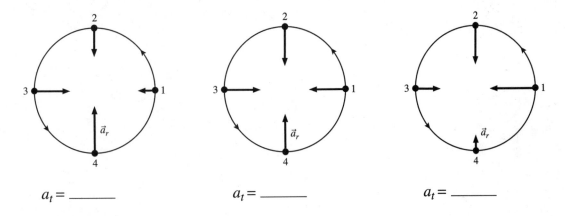

$a_t =$ _____

$a_t =$ _____

$a_t =$ _____

21. A pendulum swings from its end point on the left (point 1) to its end point on the right (point 5). At each of the labeled points:

   a. Use a **black** pen or pencil to draw and label the vectors $\vec{a}_r$ and $\vec{a}_t$ at each point. Make sure the length indicates the relative size of the vector.

   b. Use a **red** pen or pencil to draw and label the total acceleration vector $\vec{a}$.

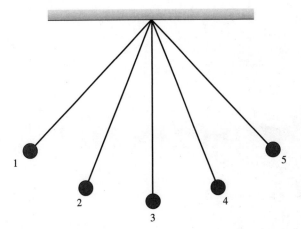

# 8 Newton's Third Law

## 8.1 Interacting Systems

## 8.2 Identifying Action/Reaction Pairs

**Exercises 1–5:**
- Draw a picture showing each relevant object *separate* from all other objects, but in the correct spatial orientation. Include the earth and, if appropriate, the earth's surface.
- Identify *all* forces and show them as **black** vectors on the objects. Label each vector as the appropriate $\vec{F}_{A\ on\ B}$.
- Connect all action/reaction pairs with **red** dotted lines.

**Note:** Your pictures should look similar to Figures 8.7 and 8.10.

1. A boy pulls a wagon by a rope attached to the front of the wagon. The rope is parallel to the ground. Rolling friction is not negligible.

2. A bicycle accelerates forward from rest. (Treat the bicycle and its rider as a single object.)

3. a. A bat hits a ball. (Draw your picture from the perspective of someone seeing the *end* of the bat at the moment it strikes the ball.)

   b. The ball then sails through the air.

4. a. A ball hangs from a string. The string is attached to the ceiling.

   b. The string is cut. The ball falls and bounces. Consider the instant that the ball is in contact with the ground. (You don't need to show the string or the ceiling in part b.)

5. A crate is in the back of a truck as the truck accelerates forward. The crate does not slip. (Treat the crate and the truck as separate systems.)

6. You find yourself in the middle of a frozen lake with a surface so slippery ($\mu_s = \mu_k = 0$) that you cannot walk. However, you happen to have several rocks in your pocket. The ice is extremely hard. It cannot be chipped, and the rocks slip on it just as much as your feet do. Can you think of a way to get to shore? Use pictures, forces, and Newton's laws to explain your reasoning.

7. How do you paddle a canoe in the forward direction? Explain. Your explanation should include pictures showing forces on the water and forces on the paddle.

8. When you blow up a balloon and release it, it shoots forward. Explain why. Include pictures showing forces on the balloon and forces on the parcel of air that was just expelled from the balloon.

9. How does a rocket take off? What is the upward force on it? Your explanation should include pictures showing forces on the rocket and forces on the parcel of hot gas that was just expelled from the rocket's exhaust.

10. How do basketball players jump straight up into the air? Your explanation should include pictures showing forces on the player and forces on the ground.

## 8.3  Newton's Third Law

11. Block A is pushed across a horizontal surface at a *constant* speed by a hand that exerts force $\vec{F}_{\text{H on A}}$. The surface has friction.

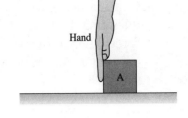

   a. Draw two free-body diagrams, one for the hand and the other for the block. On these diagrams:

   - Show only the *horizontal* forces, such as was done in Figure 8.14 of the text.
   - Label force vectors, using the form $\vec{F}_{\text{C on D}}$.
   - Connect action/reaction pairs with dotted lines.
   - On the hand diagram show only $\vec{F}_{\text{H on A}}$. Don't include $\vec{F}_{\text{body on H}}$.
   - Make sure vector lengths correctly portray the relative magnitudes of the forces.

   b. Rank in order, from largest to smallest, the magnitudes of *all* of the horizontal forces you showed in part a. For example, if $F_{\text{C on D}}$ is the largest of three forces while $F_{\text{D on C}}$ and $F_{\text{D on E}}$ are smaller but equal, you can record this as $F_{\text{C on D}} > F_{\text{D on C}} = F_{\text{D on E}}$.

   Order:

   Explanation:

   c. Repeat both part a and part b for the case that the block is *speeding up*.

12. A second block B is placed in front of Block A of question 11. B is more massive than A: $m_B > m_A$. The blocks are speeding up.

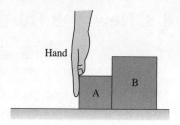

a. Consider a *frictionless* surface. Draw *separate* free-body diagrams for A, B, and the hand. Show only the horizontal forces. Label forces in the form $\vec{F}_{\text{C on D}}$. Use dotted lines to connect action/reaction pairs.

b. By applying the second law to each block and the third law to each action/reaction pair, rank in order *all* of the horizontal forces, from largest to smallest.

Order:

Explanation:

c. Repeat parts a and b if the surface has friction. Assume that A and B have the same coefficient of kinetic friction.

13. Blocks A and B are held on the palm of your outstretched hand as you lift them straight up at *constant speed*. Assume $m_B > m_A$ and that $m_{hand} = 0$.

   a. Draw *separate* free-body diagrams for A, B, and your hand.
   - Show *all* vertical forces, including the blocks' weights.
   - For your hand, show only forces exerted by the blocks; neglect the weight of your hand or any forces exerted on your hand by your arm.
   - Make sure vector lengths indicate the relative sizes of the forces.
   - Label forces in the form $\vec{F}_{C \, on \, D}$.
   - Connect action/reaction pairs with dotted lines.

   b. Rank in order, from largest to smallest, all of the vertical forces. Explain your reasoning.

14. A red car and a blue car, traveling with equal speeds, collide head-on and come to rest. How does the force the red car exerts on the blue car compare to the force that the blue car exerts on the red car? Are they equal? Is one larger than the other? Is it possible to tell? Explain.

15. A mosquito collides head-on with a car traveling 60 mph.

   a. How do you think the size of the force that the car exerts on the mosquito compares to the size of the force that the mosquito exerts on the car?

b. Draw *separate* free-body diagrams of the car and the mosquito at the moment of collision, showing only the horizontal forces. Label forces in the form $\vec{F}_{\text{C on D}}$. Connect action/reaction pairs with dotted lines.

c. Does your answer to part b confirm your answer to part a? Explain why or why not.

**Exercises 16–20:** Write the acceleration constraint in terms of *components*. For example, write $(a_1)_x = (a_2)_y$, if that is the appropriate answer, rather than $\vec{a}_1 = \vec{a}_2$.

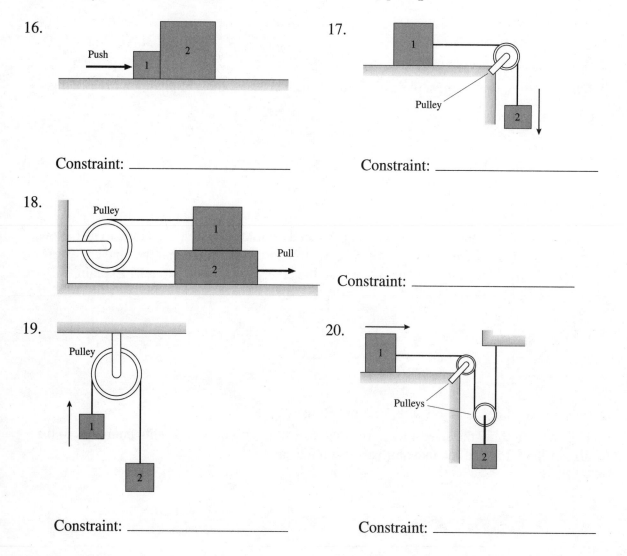

16.

Push

2

1

Constraint: _____

17.

1

Pulley

2

Constraint: _____

18.

Pulley

1

Pull

2

Constraint: _____

19.

Pulley

1

2

Constraint: _____

20.

1

Pulleys

2

Constraint: _____

# 8.4 Ropes and Pulleys

**Exercises 21–26:** Determine the reading of the spring scale.

- All the masses are at rest.
- The strings and pulleys are massless, and the pulleys are frictionless.
- The spring scale reads in kg.

21.

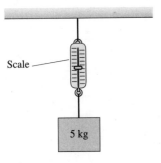

Scale = _____

22.

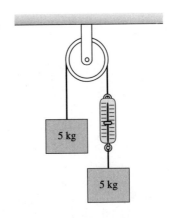

Scale = _____

23.

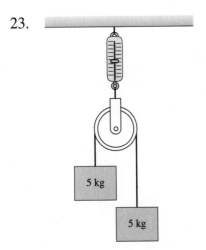

Scale = _____

24.

Scale = _____

25.

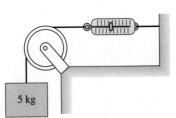

Scale = _____

26.

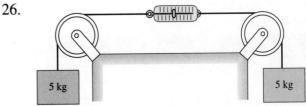

Scale = _____

27. a. A tight-rope walker at the circus steps onto the high wire, causing it to sag slightly. Is the tension in the wire less than, greater than, or equal to the performer's weight? Explain. Include a free-body diagram as part of your explanation.

    b. The leading circus magazine advertises a new wire made of a material called DreamRope. The ad says that a DreamRope wire will remain perfectly straight and horizontal, with absolutely no sag, as the performer walks across. Should you order some? Explain.

## 8.5  Examples of Interacting-System Problems

28. Blocks A and B, with $m_B > m_A$, are connected by a string. A hand pushing on the back of A accelerates them along a frictionless surface. The string (S) is massless.

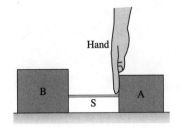

a. Draw separate free-body diagrams for A, S, and B, showing only horizontal forces. Be sure vector lengths indicate the relative size of the force. Connect any action/reaction pairs with dotted lines.

b. Rank in order, from largest to smallest, all of the horizontal forces. Explain.

c. Repeat parts a and b if the string has mass.

d. You might expect to find $F_{S \text{ on } B} > F_{H \text{ on } A}$ because $m_B > m_A$. Did you? Explain why $F_{S \text{ on } B} > F_{H \text{ on } A}$ is or is not a correct statement.

29. Blocks A and B are connected by a massless string over a massless, frictionless pulley. The blocks have just this instant been released from rest.

    a. Will the blocks accelerate? If so, in which directions?

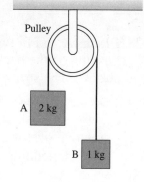

    b. Draw a separate free-body diagram for each block. Be sure vector lengths indicate the relative size of the force. Connect any action/reaction pairs or "as if" pairs with dotted lines.

    c. Rank in order, from largest to smallest, all of the vertical forces. Explain.

    d. Compare the magnitude of the *net* force on A with the *net* force on B. Are they equal, or is one larger than the other? Explain.

    e. Consider the block that falls. Is the magnitude of its acceleration less than, greater than, or equal to *g*? Explain.

30. A very smart three-year-old child is given a wagon for her birthday. She refuses to use it. "After all," she says, "Newton's third law says that no matter how hard I pull, the wagon will exert an equal but opposite force on me. So I will never be able to get it to move forward." What would you say to her in reply?

31. Will hanging a magnet in front of an iron cart make it go? Explain why or why not.

32. In case a, block A is accelerated across a frictionless table by a hanging 10 N weight (1.02 kg). In case b, the same block is accelerated by a steady 10 N tension in the string.

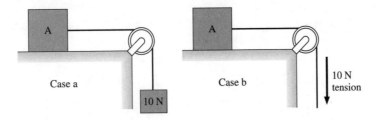

Is block A's acceleration in case b greater than, less than, or equal to its acceleration in case a? Explain.

**Exercises 33–34:** Draw separate free-body diagrams for blocks A and B. Connect any action/reaction pairs (or forces that act *as if* they are action/reaction pairs) together with dotted lines.

33.

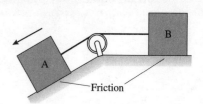

34.

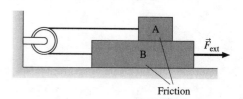

# 9 Impulse and Momentum

## 9.1 Momentum

## 9.2 Solving Impulse and Momentum Problems

1. Rank in order, from largest to smallest, the momenta $(p_x)_1$ to $(p_x)_5$.

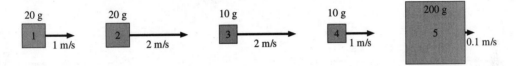

Order:

2. The position-versus-time graph is shown for a 500 g object. Draw the corresponding momentum-versus-time graph.

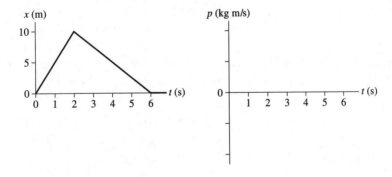

3. The momentum-versus-time graph is shown for a 500 g object. Draw the corresponding acceleration-versus-time graph.

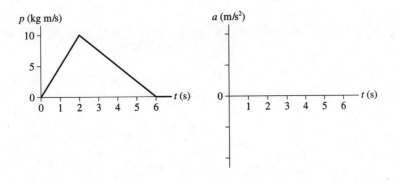

4. Explain the concept of *impulse* in nonmathematical language. That is, don't simply put the equation in words to say that "impulse is the time integral of force." Explain it in terms that would make sense to an educated person who had never heard of it.

5. A 2 kg object is moving to the right with a speed of 1 m/s when it experiences an impulse due to the force shown in the graph. What is the object's speed and direction after the impulse?

a.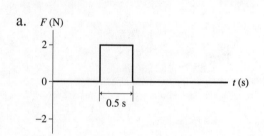

b.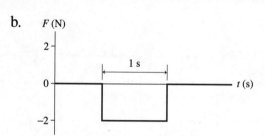

6. A 2 kg object is moving to the right with a speed of 1 m/s when it experiences an impulse due to the force shown in the graph. What is the object's speed and direction after the impulse?

a.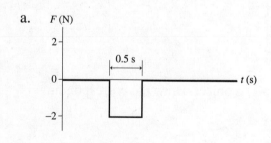

b.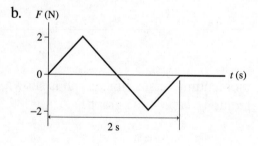

7. A 0.2 kg plastic cart and a 20 kg lead cart both roll without friction on a horizontal surface. Equal forces are used to push both carts forward for a time of 1 s, starting from rest. After the force is removed at $t = 1$ s, is the momentum of the plastic cart greater than, less than, or equal to the momentum of the lead cart? Explain.

8. A carnival game requires you to knock over a wood post by throwing a ball at it. You're offered a very bouncy rubber ball and a very sticky clay ball of equal mass. Assume that you can throw them with equal speed and equal accuracy. You only get one throw.

a. Which ball will you choose? Why?

b. Let's think about the situation more carefully. Both balls have the same initial momentum $p_{ix}$ just before hitting the post. The clay ball sticks, the rubber ball bounces off with essentially no loss of speed. What is the final momentum of each ball?

Clay ball: $p_{fx} =$ _____          Rubber ball $p_{fx} =$ _____

Hint: Momentum has a sign. Did you take the sign into account?

c. What is the *change* in the momentum of each ball?

Clay ball: $\Delta p_x =$ _____          Rubber ball $\Delta p_x =$ _____

d. Which ball experiences a larger impulse during the collision? Explain.

e. From Newton's third law, the impulse that the ball exerts on the post is equal in magnitude, although opposite in direction, to the impulse that the post exerts on the ball. Which ball exerts the larger impulse on the post?

f. Don't change your answer to part a, but are you still happy with that answer? If not, how would you change your answer? Why?

9. For each of the following situations, use both words and pictures to
   - Describe what happens in the language of force, acceleration, and action/reaction.
   - Describe what happens in the language of impulse and momentum.

   a. A moving blob of clay hits a stationary bowling ball.

   Force description:

   Momentum description:

   b. A falling rubber ball bounces off the floor.

   Force description:

   Momentum description:

   c. Two equal masses are pushed apart by a compressed spring between them.

   Force description:

   Momentum description:

10. A small, light ball S and a large, heavy ball L move toward each other, collide, and bounce apart.

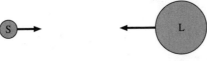

a. Compare the force that S exerts on L to the force that L exerts on S. That is, is $F_{S \text{ on } L}$ larger, smaller, or equal to $F_{L \text{ on } S}$? Explain.

b. Compare the time interval during which S experiences a force to the time interval during which L experiences a force. Are they equal, or is one longer than the other?

c. Sketch a graph showing a *plausible* $F_{S \text{ on } L}$ as a function of time and another graph showing a plausible $F_{L \text{ on } S}$ as a function of time. Be sure think about the *sign* of each force.

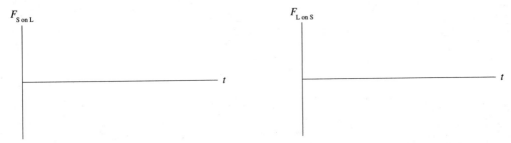

d. Compare the impulse delivered to S to the impulse delivered to L. Explain.

e. Compare the momentum change of S to the momentum change of L.

f. Compare the velocity change of S to the velocity change of L.

g. What is the change in the *sum* of the momenta of the two balls? Is it positive, negative, or zero?

**Exercises 11–13:** Draw a momentum bar chart to show the momenta and impulse for the situation described.

11. A compressed spring shoots a ball to the right. The ball was initially at rest.

$$+ \quad \begin{array}{ccccc} p_{ix} & + & J_x & = & p_{fx} \end{array}$$

12. A rubber ball is tossed straight up and bounces off the ceiling. Consider only the collision with the ceiling.

$$+ \quad \begin{array}{ccccc} p_{ix} & + & J_x & = & p_{fx} \end{array}$$

13. A clay ball is tossed straight up and sticks to the ceiling. Consider only the collision with the ceiling.

$$+ \quad \begin{array}{ccccc} p_{ix} & + & J_x & = & p_{fx} \end{array}$$

## 9.3 Conservation of Momentum

14. Explain the concept of "isolated system" in nonmathematical language.

15. A golf club continues forward after hitting the golf ball. Is momentum conserved in the collision? Explain, making sure you are careful to identify the "system."

16. As you release a ball, it falls—gaining speed and momentum. Is momentum conserved?
    a. Answer this question from the perspective of choosing the ball alone as the system.

    b. Answer this question from the perspective of choosing ball + earth as the system.

17. Two particles collide, one of which was initially moving and the other initially at rest.
    a. Is it possible for *both* particles to be at rest after the collision? Give an example in which this happens, or explain why it can't happen.

    b. Is it possible for *one* particle to be at rest after the collision? Give an example in which this happens, or explain why it can't happen.

## 9.4 Explosions

## 9.5 Inelastic Collisions

**Exercises 18–20:** Prepare a pictorial representation for these problems, but *do not* solve them.
- Draw pictures of "before" and "after."
- Define symbols relevant to the problem.
- List known information, and identify the desired unknown.

18. A 50 kg archer, standing on frictionless ice, shoots a 100 g arrow at a speed of 100 m/s. What is the recoil speed of the archer?

19. The parking brake on a 2000 kg Cadillac has failed, and it is rolling slowly, at 1 mph, toward a group of small innocent children. As you see the situation, you realize there is just time for you to drive your 1000 kg Volkswagen head-on into the Cadillac and thus to save the children. With what speed should you impact the Cadillac to bring it to a halt?

20. Dan is gliding on his skateboard at 4 m/s. He suddenly jumps backward off the skateboard, kicking the skateboard forward at 8 m/s. How fast is Dan going as his feet hit the ground? Dan's mass is 50 kg and the skateboard's mass is 5 kg.

# 9.6 Momentum in Two Dimensions

21. An object initially at rest explodes into three fragments. The momentum vectors of two of the fragments are shown. Draw the momentum vector $\vec{p}_3$ of the third fragment.

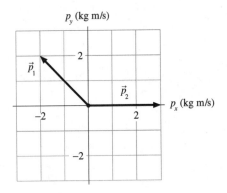

22. An object initially at rest explodes into three fragments. The momentum vectors of two of the fragments are shown. Draw the momentum vector $\vec{p}_3$ of the third fragment.

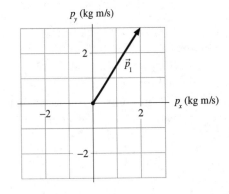

23. A 500 g ball traveling to the right at 4.0 m/s collides with and bounces off another ball. The figure shows the momentum vector $\vec{p}_1$ of the first ball after the collision. Draw the momentum vector $\vec{p}_2$ of the second ball.

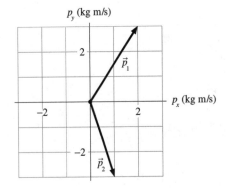

24. A 500 g ball traveling to the right at 4.0 m/s collides with and bounces off another ball. The figure shows the momentum vector $\vec{p}_1$ of the first ball after the collision. Draw the momentum vector $\vec{p}_2$ of the second ball.

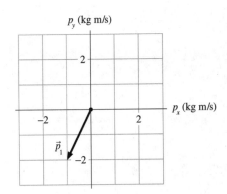

## 9.7 Angular Momentum

25. Rank in order, from largest to smallest, the angular momenta $L_1$ to $L_5$ of these particles.

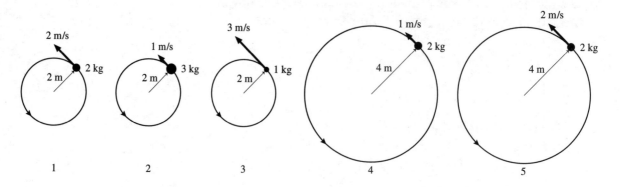

1      2      3      4      5

Order:

Explanation:

26. Two masses are held together by a thread on a rod that is rotating about the center with angular velocity $\omega$. If the thread breaks, the masses will slide out to the ends of the rod. If that happens, will the rod's angular velocity increase, decrease, or remain unchanged? Explain.

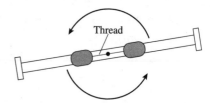

# 10 Energy

## 10.1 A "Natural Money" Called Energy

1. One month, John has income of $3000, expenses of $2500, and he sells $300 of stocks.

   a. Can you determine John's liquid assets $L$ at the end of the month? If so, what is $L$? If not, why not?

   b. Can you determine the amount by which John's liquid assets *changed* during the month? If so, what is $\Delta L$?

2. John begins the month with $2000 of liquid assets and $5000 of savings. His financial activity for the month is as follows:

   | Day of Month | Activity |
   | --- | --- |
   | 1 | Receives a $3000 paycheck; deposits it in checking |
   | 3 | Spends $500 |
   | 8 | Buys a $1000 savings bond |
   | 10 | Pays bills totaling $1000 |
   | 15 | Receives a $100 birthday present from Grandma |
   | 23 | Sells $1500 of stock |
   | 28 | Buys a $1200 bicycle |

   a. What are John's liquid assets and saved assets at the end of the month?

   b. Show that John's monetary relationship $\Delta W = I - E$ is satisfied.

## 10.2 Kinetic Energy and Gravitational Potential Energy

3. Upon what basic quantity does kinetic energy depend? _____

   Upon what basic quantity does potential energy depend? _____

4. Can kinetic energy ever be negative? _____
   Give a plausible *reason* for your answer without making use of any formulas.

5. Can gravitational potential energy ever be negative? _____
   Give a plausible *reason* for your answer without making use of any formulas.

6. a. If a particle's speed increases by a factor of three, by what factor does its kinetic energy change?

   b. Particle A has half the mass and eight times the kinetic energy of particle B. What is the speed ratio $v_A/v_B$?

7. On the axes below, draw graphs of the kinetic energy of
   a. A 1000 kg car that uniformly accelerates from 0 to 20 m/s in 20 s.
   b. A 1000 kg car moving at 20 m/s that brakes to a halt with uniform deceleration in 4 s.
   c. A 1000 kg car that drives once around a 40-m-diameter circle at a speed of 20 m/s.
   Calculate $K$ at several times, plot the points, and draw a smooth curve between them.

a.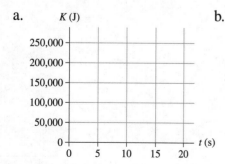

b.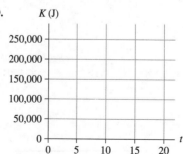

c.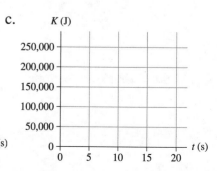

## 10.3  A Closer Look at Gravitational Potential Energy

8. Below we see a 1 kg object that is initially 1 m above the ground and rises to a height of 2 m. Anjay, Brittany, and Carlos each measure its position, but each of them uses a different coordinate system. Fill in the table to show the initial and final gravitational potential energies and $\Delta U$ as measured by our three aspiring scientists.

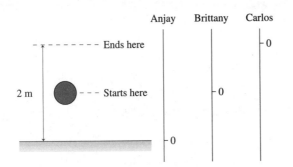

|  | $U_i$ | $U_f$ | $\Delta U$ |
|---|---|---|---|
| Anjay |  |  |  |
| Brittany |  |  |  |
| Carlos |  |  |  |

9. A roller coaster car rolls down a frictionless track, reaching speed $v_f$ at the bottom.

   a. If you want the car to go twice as fast at the bottom, by what factor must you increase the height of the track?

   b. Does your answer to part a depend on whether the track is straight or not? Explain.

10. Three balls of equal mass are fired simultaneously with *equal* speeds from the same height above the ground. Ball 1 is fired straight up, ball 2 is fired straight down, and ball 3 is fired horizontally. Rank in order, from largest to smallest, their speeds $v_1$, $v_2$, and $v_3$ as they hit the ground.

   Order:

   Explanation:

11. Below are shown three frictionless tracks. A ball is released from rest at the position shown on the left. To which point does the ball make it on the right before reversing direction and rolling back? Point B is the same height as the starting position.

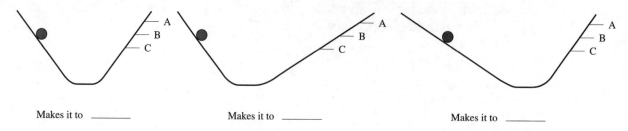

Makes it to _____          Makes it to _____          Makes it to _____

**Exercises 12–14:** Draw an energy bar chart to show the energy transformations for the situation described.

12. A car runs out of gas and coasts up a hill until finally stopping.

$$K_i \quad + \quad U_{gi} \quad = \quad K_f \quad + \quad U_{gf}$$

13. A pendulum is held out at 45° and released from rest. A short time later it swings through the lowest point on its arc.

$$K_i \quad + \quad U_{gi} \quad = \quad K_f \quad + \quad U_{gf}$$

14. A ball starts from rest on the top of one hill, rolls without friction through a valley, and just barely makes it to the top of an adjacent hill.

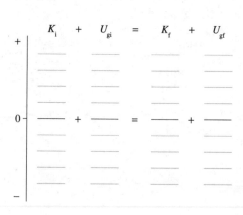

## 10.4  Restoring Forces and Hooke's Law

15. A spring is attached to the floor and pulled straight up by a string. The spring's tension is measured. The graph shows the tension in the spring as a function of the spring's length $L$.

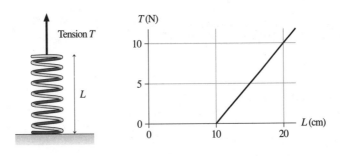

   a. Does this spring obey Hooke's Law? Explain why or why not.

   b. If it does, what is the spring constant?

16. Draw a figure analogous to Figure 10.17 in the textbook for a spring that is attached to a wall on the *right* end. Use the figure to show that $F$ and $\Delta s$ always have opposite signs.

17. A spring has an unstretched length of 10 cm. It exerts a restoring force $F$ when stretched to a length of 11 cm.

    a. For what length of the spring is its restoring force $3F$?

    b. At what compressed length is the restoring force $2F$?

18. The left end of a spring is attached to a wall. When Bob pulls on the right end with a 200 N force, he stretches the spring by 20 cm. The same spring is then used for a tug-of-war between Bob and Carlos. Each pulls on his end of the spring with a 200 N force.

    a. How far does Bob's end of the spring move? Explain.

    b. How far does Carlos's end of the spring move? Explain.

## 10.5 Elastic Potential Energy

19. A heavy object is released from rest at position 1 above a spring. It falls and contacts the spring at position 2. The spring achieves maxiumum compression at position 3. Fill in the table below to indicate whether each of the quantities are +, −, or 0 during the intervals 1→2, 2→3, and 1→3.

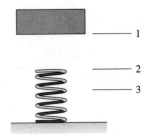

|  | 1→2 | 2→3 | 1→3 |
|---|---|---|---|
| $\Delta K$ |  |  |  |
| $\Delta U_g$ |  |  |  |
| $\Delta U_s$ |  |  |  |

20. Rank in order, from most to least, the amount of elastic potential energy $(U_s)_1$ to $(U_s)_4$ stored in each of these springs.

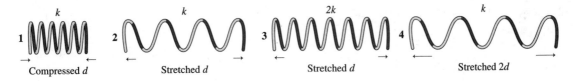

| $k$ | $k$ | $2k$ | $k$ |
|---|---|---|---|
| 1 | 2 | 3 | 4 |
| Compressed $d$ | Stretched $d$ | Stretched $d$ | Stretched $2d$ |

Order:

Explanation:

21. A spring gun shoots out a plastic ball at speed $v_0$. The spring is then compressed twice the distance it was on the first shot.

a. By what factor is the spring's potential energy increased?

b. By what factor is the ball's velocity increased? Explain.

**Exercises 22–23:** Draw an energy bar chart to show the energy transformations for the situation described.

22. A bobsled sliding across frictionless, horizontal ice runs into a giant spring. A short time later the spring reaches its maximum compression.

$$K_i + U_{gi} + U_{si} = K_f + U_{gf} + U_{sf}$$

23. A brick is held above a spring that is standing on the ground. The brick is released from rest, and a short time later the spring reaches its maximum compression.

$$K_i + U_{gi} + U_{si} = K_f + U_{gf} + U_{sf}$$

## 10.6 Elastic Collisions

24. Ball 1 with an initial speed of 14 m/s has a perfectly elastic collision with ball 2 that is initially at rest. Afterward, the speed of ball 2 is 21 m/s.

a. What will be the speed of ball 2 if the initial speed of ball 1 is doubled?

b. What will be the speed of ball 2 if the mass of ball 1 is doubled?

## 10.7 Energy Diagrams

25. The figure shows a potential-energy curve. Suppose a particle with total energy $E_1$ is at position A and moving to the right.

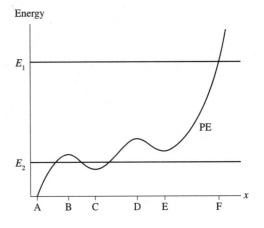

a. For each of the following regions of the $x$-axis, does the particle speed up, slow down, maintain a steady speed, or change direction?

A to B  _____

B to C  _____

C to D  _____

D to E  _____

E to F  _____

b. Where is the particle's turning point?  _____

c. For a particle that has total energy $E_2$, what are the possible motions and where do they occur along the $x$-axis?

d. What position or positions are points of stable equilibrium? For each, would a particle in equilibrium at that point have total energy $\le E_2$, between $E_2$ and $E_1$, or $\ge E_1$?

e. What position or positions are points of unstable equilibrium? For each, would a particle in equilibrium at that point have total energy $\le E_2$, between $E_2$ and $E_1$, or $\ge E_1$?

26. A particle with the potential energy shown in the graph is moving to the right at $x = 0$ m with total energy $E$.

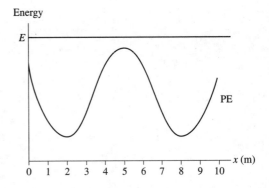

a. At what value or values of $x$ is the particle's speed a maximum?

b. At what value or values of $x$ is the particle's speed a minimum?

c. At what value or values of $x$ is the potential energy a maximum?

d. Does this particle have a turning point in the range of $x$ covered by the graph? If so, where?

27. Below are a set of axes on which you are going to draw a potential-energy curve. By doing experiments, you find the following information:
    - A particle with energy $E_1$ oscillates between positions D and E.
    - A particle with energy $E_2$ oscillates between positions C and F.
    - A particle with energy $E_3$ oscillates between positions B and G.
    - A particle with energy $E_4$ enters from the right, bounces at A, then never returns.

    Draw a potential-energy curve that is consistent with this information.

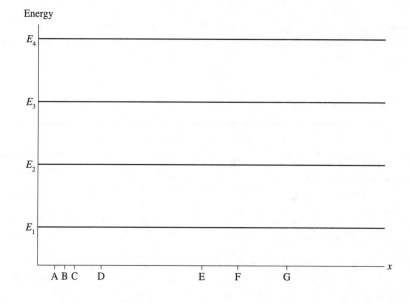

# 11 Work

## 11.1 The Basic Energy Model

1. What are the two primary processes by which energy can be transferred from the environment to a system?

2. a. A process occurs in which a system's potential energy decreases while the environment does work on the system. Does the system's kinetic energy increase, decrease, or stay the same? Or is there not enough information to tell? Explain.

   b. A process occurs in which a system's potential energy increases while the environment does work on the system. Does the system's kinetic energy increase, decrease, or stay the same? Or is there not enough information to tell? Explain.

3. The kinetic energy of a system decreases and its potential energy is unchanged. What is doing work on what? That is, does the environment do work on the system, or does the system do work on the environment? Explain.

## 11.2  Work and Kinetic Energy

## 11.3  Calculating and Using Work

4. For each situation described below:
   - Draw a before-and-after diagram, similar to Figures 11.7 and 11.10 in the textbook.
   - Identify *all* forces acting on the particle.
   - Determine if the work done by each of these forces is positive (+), negative (−), or zero (0). Make a little table beside the figure showing *every* force and the sign of its work.

   a. An elevator moves upward.

   b. An elevator moves downward.

   c. You push a box across a rough floor.

d. You slide down a steep hill.

e. A ball is thrown straight up. Consider the ball from one microsecond after it leaves your hand until the highest point of its trajectory.

f. A car turns a corner at constant speed.

5. For each pair of vectors, is the sign of $\vec{A} \cdot \vec{B}$ positive (+), negative (−), or zero (0)?

a.
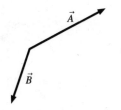

Sign = _____

b.

Sign = _____

c.

Sign = _____

d.
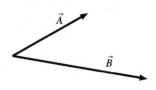

Sign = _____

e.
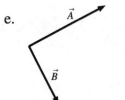

Sign = _____

f.
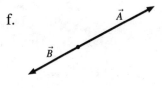

Sign = _____

6. Each of the diagrams below shows a vector $\vec{A}$. Draw and label a vector $\vec{B}$ that will cause $\vec{A} \cdot \vec{B}$ to have the sign indicated.

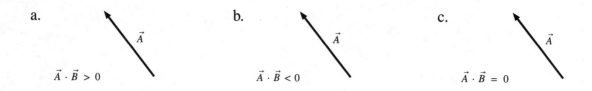

a.

$\vec{A}$

$\vec{A} \cdot \vec{B} > 0$

b.

$\vec{A}$

$\vec{A} \cdot \vec{B} < 0$

c.

$\vec{A}$

$\vec{A} \cdot \vec{B} = 0$

7. For each situation described below:
   - Draw a before-and-after diagram, similar to Figures 11.7 and 11.10 in the textbook.
   - Draw and label the displacement vector $\Delta\vec{r}$.
   - Draw a free-body diagram showing all forces on the object.
   - Make a little table to show the sign (+, −, or 0) of
     i.   $W$ for each force on the free-body diagram,
     ii.  $W_{net}$, and
     iii. $\Delta K$.

   a. A box slides to a stop along a horizontal floor with friction.

   b. A box slides down a frictionless slope.

   c. A box slides up a frictionless slope.

d. A ball rises after being tossed straight up.

e. A descending elevator brakes to a halt.

8. A 0.5 kg mass on a 1-m-long string swings in a circle on a horizontal, frictionless table at a steady speed of 2 m/s.

a. How much work does the tension in the string do on the mass during one revolution? Explain.

b. Is your answer to part a consistent with the work-kinetic energy theorem? Explain.

9. A 0.2 kg plastic cart and a 20 kg lead cart both roll without friction on a horizontal surface. Equal forces are used to push both carts forward a distance of 1 m, starting from rest. After traveling 1 m, is the kinetic energy of the plastic cart greater than, less than, or equal to the kinetic energy of the lead cart? Explain.

10. Particle A has less mass than particle B. Both are pushed forward across a frictionless surface by equal forces for 1 s. Both start from rest.

   a. Compare the amount of work done on each particle. That is, is the work done on A greater than, less than, or equal to the work done on B? Explain.

   b. Compare the impulses delivered to particles A and B. Explain.

   c. Compare the final speeds of particles A and B. Explain.

## 11.4 The Work Done by a Variable Force

11. In Chapter 9, we found a graphical interpretation of $\Delta p$ as the area under the $F$-versus-$t$ graph from an initial time $t_i$ to a final time $t_f$. Provide an analogous graphical interpretation of $\Delta K$, the change in kinetic energy.

12. A particle moving along the $x$-axis experiences the forces shown below. How much work does each force do on the particle? What is each particle's change in kinetic energy?

a.

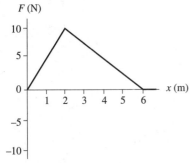

b.

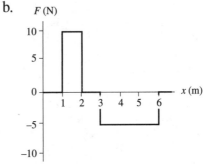

W = _____        W = _____

$\Delta K$ = _____        $\Delta K$ = _____

13. A 1 kg particle moving along the $x$-axis experiences the force shown in the graph. If the particle's speed is 2 m/s at $x = 0$ m, what is its speed when it gets to $x = 5$ m?

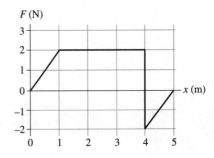

14. In Example 11.8 in the textbook, a compressed spring with a spring constant of 20 N/m expands from $x_0 = -20$ cm $= -0.20$ m to its equilibrium position at $x_1 = 0$ m.

   a. Graph the spring force $F_{sp}$ from $x_1 = -0.20$ m to $x_2 = 0$ m.

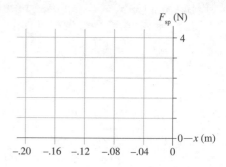

   b. Suppose the surface had been frictionless. Use your graph to determine $\Delta K$, the change in a cube's kinetic energy when launched by a spring that has been compressed by 20 cm.

   c. Use your result from part b to find the launch speed of a 100 g cube in the absence of friction. Compare your answer to the value found in the Example 11.8. Why are they different?

## 11.5  Force, Work, and Potential Energy

## 11.6  Finding Force from Potential Energy

15. A particle moves in a vertical plane along a *closed* path, starting at A and eventually returning to its starting point. How much work is done on the particle by gravity? Explain.

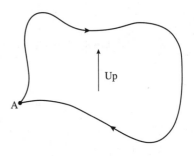

16. a. If the force on a particle at some point in space is zero, must its potential energy also be zero at that point? Explain.

    b. If the potential energy of a particle at some point in space is zero, must the force on it also be zero at that point? Explain.

17. The graph shows the potential-energy curve of a particle moving along the $x$-axis under the influence of a conservative force.

    a. In which intervals of $x$ is the force on the particle to the right?

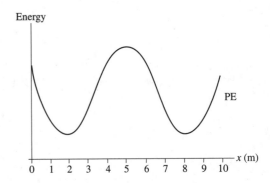

    b. In which intervals of $x$ is the force on the particle to the left?

    c. At what value or values of $x$ is the magnitude of the force a maximum?

d. What value or values of x are positions of stable equilibrium?

e. What value or values of x are positions of unstable equilibrium?

f. If the particle is released from rest at $x = 0$ m, will it reach $x = 10$ m? Explain.

## 11.7 Thermal Energy

18. A car traveling at 60 mph slams on its brakes and skids to a halt. What happened to the kinetic energy the car had just before stopping?

19. What energy transformations occur as a skier glides down a gentle slope at constant speed?

# 11.8  Conservation of Energy

20. What is meant by an *isolated system*?

21. Give a *specific* example of a situation in which:

a. $W_{ext} \rightarrow K$ with $\Delta U = 0$ and $\Delta E_{th} = 0$.

b. $W_{ext} \rightarrow U$ with $\Delta K = 0$ and $\Delta E_{th} = 0$.

c. $K \rightarrow U$ with $W_{ext} = 0$ and $\Delta E_{th} = 0$.

d. $W_{ext} \rightarrow E_{th}$ with $\Delta K = 0$ and $\Delta U = 0$.

e. $U \rightarrow E_{th}$ with $\Delta K = 0$ and $W_{ext} = 0$.

22. A system loses 1000 J of potential energy. In the process, it does 500 J of work on the environment and the thermal energy increases by 250 J. Show this process on an energy bar chart.

$$K_i \; + \; U_i \; + \; W_{ext} \; = \; K_f \; + \; U_f \; + \; \Delta E_{th}$$

23. A system gains 1000 J of kinetic energy while losing 500 J of potential energy. The thermal energy increases by 250 J. Show this process on an energy bar chart.

$$K_i \; + \; U_i \; + \; W_{ext} \; = \; K_f \; + \; U_f \; + \; \Delta E_{th}$$

24. In the text, we have found both the work-kinetic energy theorem $W = \Delta K$ and the definition of potential energy $\Delta U = -W$. What is the relationship between these two statements? Do they say the same thing, or something different? Explain.

## 11.9 Power

25. a. If you push an object 10 m with a 10 N force in the direction of motion, how much work do you do on it?

b. How much power must you provide to push the object in 1 s? In 10 s? In 0.1 s?

# 12 Newton's Theory of Gravity

## 12.1 A Little History

## 12.2 Isaac Newton

## 12.3 Newton's Law of Gravity

1. Is the earth's gravitational force on the sun larger than, smaller than, or equal to the sun's gravitational force on the earth? Explain.

2. Star A is twice as massive as star B.
   a. Draw gravitational force vectors on both stars. The length of each vector should be proportional to the size of the force.

   A
   $m_A = 2m_B$

   B
   $m_B$

   b. Is the acceleration of star A larger than, smaller than, or equal to the acceleration of star B? Explain.

3. The gravitational force of a star on orbiting planet 1 is $F_1$. Planet 2, which is twice as massive as planet 1 and orbits at twice the distance from the star, experiences gravitational force $F_2$. What is the ratio $F_2/F_1$?

4. Comets orbit the sun in highly elliptical orbits. A new comet is sighted at time $t_1$.

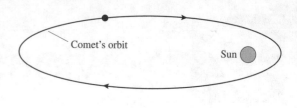

Comet's orbit

Sun

a. Later, at time $t_2$, the comet's acceleration $a_2$ is twice as large as the acceleration $a_1$ it had at $t_1$. What is the ratio $r_2/r_1$ of the comet's distance from the sun at $t_2$ to its distance at $t_1$?

b. Still later, at time $t_3$, the comet has rounded the sun and is headed back out to the farthest reaches of the solar system. The size of the force $F_3$ on the comet at $t_3$ is the same as the size of force $F_2$ at $t_2$, but the comet's distance from the sun $r_3$ is only 90% of distance $r_2$. Astronomers recognize that the comet has lost mass. Part of it was "boiled away" by the heat of the sun during the time of closest approach, thus forming the comet's tail. What percent of its initial mass did the comet lose?

## 12.4 Little *g* and Big *G*

5. How far away from the earth does an orbiting spacecraft have to be in order for the astronauts inside to be "weightless"?

6. The acceleration due to gravity at the surface of planet 1 is 20 m/s². The radius and the mass of planet 2 are twice those of planet 1. What is *g* on planet 2?

## 12.5  Gravitational Potential Energy

7. Explain *why* the gravitational potential energy of two masses is negative. Note that saying "because that's what the formula gives" is *not* an explanation. An explanation makes use of the basic ideas of force and potential energy.

## 12.6  Satellite Orbits and Energies

8. Planet X orbits the star Omega with a "year" that is 200 earth days long. Planet Y circles Omega at four times the distance of planet X. How long is a year on planet Y?

9. The mass of Jupiter is $M_{Jupiter} = 300M_{earth}$. Jupiter orbits around the sun with $T_{Jupiter} = 11.9$ years in an orbit with $r_{Jupiter} = 5.2r_{earth}$. Suppose the earth could be moved to the distance of Jupiter and placed in a circular orbit around the sun. The new period of the earth's orbit would be

a. 1 year.

b. 11.9 years.

c. Between 1 year and 11.9 years.

d. More than 11.9 years.

e. It could be anything, depending on the speed the earth is given.

f. It is impossible for a planet of earth's mass to orbit at the distance of Jupiter.

Circle the letter of the true statement. Then explain your choice.

10. Satellite A orbits a planet with a speed of 10,000 m/s. Satellite B is twice as massive as satellite A and orbits at twice the distance from the center of the planet. What is the speed of satellite B?

11. a. A crew of a spacecraft in a clockwise circular orbit around the moon wants to change to a new orbit that will take them down to the surface. In which direction should they fire the rocket engine? On the figure, show the exhaust gases coming out of the spacecraft.

    b. On the figure, show the spacecraft's orbit after firing its rocket engine.

    c. The moon has no atmosphere, so the spacecraft will continue unimpeded along its new orbit until either firing its rocket again or (ouch!) intersecting the surface. As it descends, does its speed increase, decrease, or stay the same? Explain.

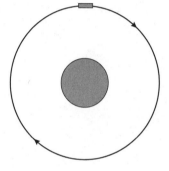

# 13 Rotation of a Rigid Body

## 13.1 Rotational Kinematics

1. The following figures show a rotating wheel. Determine the signs (+ or −) of $\omega$ and $\alpha$.

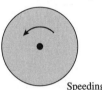
Speeding up

$\omega$ _____

$\alpha$ _____

Slowing down

$\omega$ _____

$\alpha$ _____

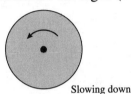
Slowing down

$\omega$ _____

$\alpha$ _____

Speeding up

$\omega$ _____

$\alpha$ _____

2. The figure shows a pendulum at one end point of its arc.

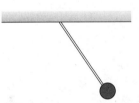

a. At this point, is $\omega$ positive, negative, or zero? _____

b. At this point, is $\alpha$ positive, negative, or zero? _____

3. The figures below show the radial acceleration vector $\vec{a}_r$ at four successive points on the trajectory of a particle moving in a counterclockwise circle.

a. For each, draw the tangential acceleration vector $\vec{a}_t$ at points 2 and 3 or, if appropriate, write $\vec{a}_t = \vec{0}$.

b. Determine if the particle's angular acceleration $\alpha$ is positive (+), negative (−), or zero (0).

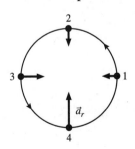

$\alpha =$ _____

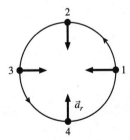

$\alpha =$ _____

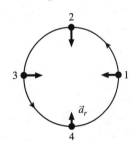

$\alpha =$ _____

13-1

4. The figure shows the $\theta$-versus-$t$ graph for a particle moving in a circle. The curves are all sections of parabolas.

   a. Draw the corresponding $\omega$-versus-$t$ and $\alpha$-versus-$t$ graphs. Notice that the horizontal tick marks are equally spaced.

   b. Write a description of the particle's motion.

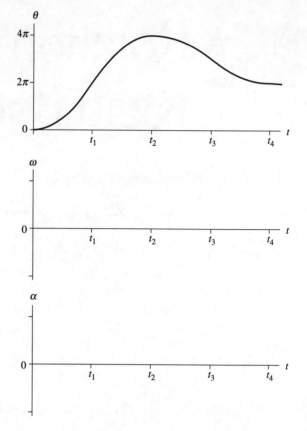

5. A wheel rolls to the left along a horizontal surface, up a ramp, then continues along the upper horizontal surface. Draw graphs for the wheel's angular velocity $\omega$ and angular acceleration $\alpha$ as a functions of time.

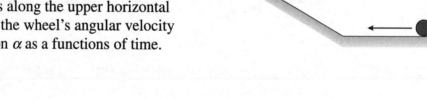

## 13.2  Rotation About the Center of Mass

6. Is the center of mass of this dumbbell at point 1, 2, or 3? Explain.

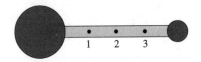

7. Mark the center of mass of this object with an ×.

## 13.3  Torque

8. Five forces are applied to a door. For each, determine if the torque about the hinge is positive (+), negative (−), or zero (0).

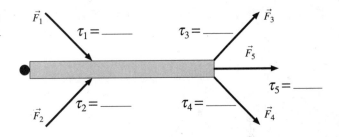

9. Six forces, each of magnitude either $F$ or $2F$, are applied to a door. Rank in order, from largest to smallest, the six torques $\tau_1$ to $\tau_6$ about the hinge.

Order:

Explanation:

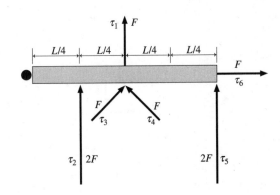

10. Four forces are applied to a rod that can pivot on an axle. For each force,

a. Use a **black** pen or pencil to draw the line of action.

b. Use a **red** pen or pencil to draw and label the moment arm, or state that $d = 0$.

c. Determine if the torque about the axle is positive (+), negative (−), or zero (0).

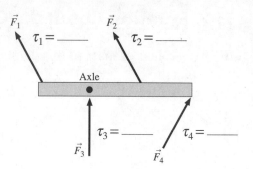

11. a. Draw a force vector at A whose torque about the axle is negative.

b. Draw a force vector at B whose torque about the axle is zero.

c. Draw a force vector at C whose torque about the axle is positive.

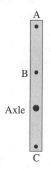

12. a. Draw a second force $\vec{F}_2$ that forms a couple with $\vec{F}_1$.

b. Draw and label the distance $l$ between their lines of action.

c. Is the torque positive, negative, or zero? Explain.

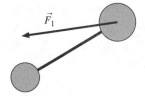

13. The dumbbells below are all the same size, and the forces all have the same magnitude. Rank in order, from largest to smallest, the torques $\tau_1$, $\tau_2$, and $\tau_3$.

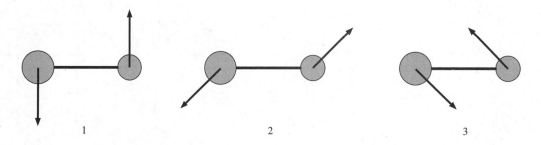

Order:

Explanation:

## 13.4  Rotational Dynamics

14. A student gives a quick push to a ball at the end of a massless, rigid rod, causing the ball to rotate clockwise in a *horizontal* circle. The rod's pivot is frictionless.

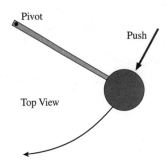

    a. As the student is pushing, is the torque about the pivot positive, negative, or zero?

    b. After the push has ended, does the ball's angular velocity
       i.    Steadily increase?
       ii.   Increase for awhile, then hold steady?
       iii.  Hold steady?
       iv.   Decrease for awhile, then hold steady?
       v.    Steadily decrease?
       Explain the reason for your choice.

    c. Right after the push has ended, is the torque positive, negative, or zero? _____

15. a. Rank in order, from largest to smallest, the torques $\tau_1$ to $\tau_4$.

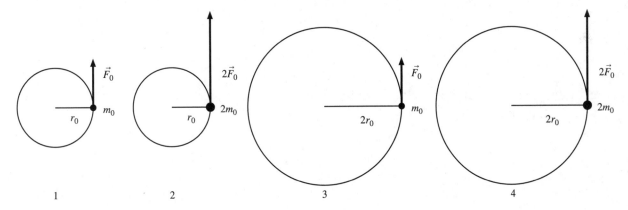

    Order:

    Explanation:

    b. Rank in order, from largest to smallest, the angular accelerations $\alpha_1$ to $\alpha_4$.

16. The top graph shows the torque on a rotating wheel as a function of time. The wheel's moment of inertia is 10 kg m$^2$. Draw graphs of $\alpha$-versus-$t$ and $\omega$-versus-$t$, assuming $\omega_0 = 0$.

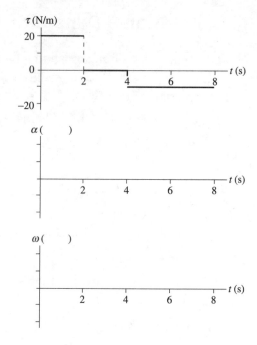

17. The wheel turns on a frictionless axle. A string wrapped around the smaller diameter shaft is tied to a block. The block is released at $t = 0$ s and hits the ground at $t = t_1$.

    a. Draw a graph of $\omega$-versus-$t$ for the wheel, starting at $t = 0$ s and continuing to some time $t > t_1$.

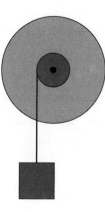

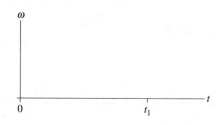

    b. Is the magnitude of the block's downward acceleration greater than $g$, less than $g$, or equal to $g$? Explain.

18. The moment of inertia of a uniform rod about an axis through its center is $\frac{1}{12}ML^2$. The moment of inertia about an axis at one end is $\frac{1}{3}ML^2$. Explain *why* the moment of inertia is larger about the end than about the center.

19. You have two steel spheres. Sphere 2 has twice the radius of sphere 1. By what *factor* does the moment of inertia $I_2$ of sphere 2 exceed the moment of inertia $I_1$ of sphere 1?

20. The professor hands you two spheres. They have the same mass, the same radius, and the same exterior surface. The professor claims that one is a solid sphere and that the other is hollow. Can you determine which is which without cutting them open? If so, how? If not, why not?

21. Rank in order, from largest to smallest, the moments of inertia $I_1$, $I_2$, and $I_3$.

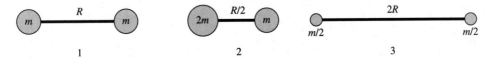

Order:

Explanation:

# 13.5  Rotation About a Fixed Axis

22. A square plate can rotate about an axle through its center. Four forces of equal magnitude are applied to different points on the plate. The forces turn as the plate rotates, maintaining the same orientation with respect to the plate. Rank in order, from largest to smallest, the angular accelerations $\alpha_1$ to $\alpha_4$.

    Order:

    Explanation:

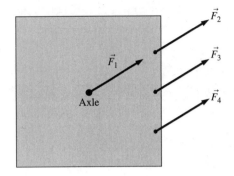

23. A solid cylinder and a cylindrical shell have the same mass, same radius, and turn on frictionless, horizontal axles. (The cylindrical shell has light-weight spokes connecting the shell to the axle.) A rope is wrapped around each cylinder and tied to a block. The blocks have the same mass and are held the same height above the ground. Both blocks are released simultaneously. The ropes do not slip.

    Which block hits the ground first? Or is it a tie? Explain.

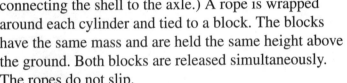

# 13.6  Rigid Body Equilibrium

24. A uniform rod pivots about a frictionless, horizontal axle through its center. It is placed on a stand, held motionless in the position shown, then gently released. On the right side of the figure, draw the final, equilibrium position of the rod. Explain your reasoning.

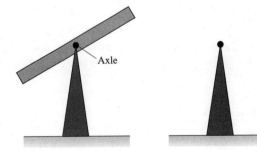

25. The dumbbell has masses $m$ and $2m$. Force $\vec{F}_1$ acts on mass $m$ in the direction shown. Is there a force $\vec{F}_2$ that can act on mass $2m$ such that the dumbbell moves with pure translational motion, without any rotation? If so, draw $\vec{F}_2$, making sure that its length shows the magnitude of $\vec{F}_2$ relative to $\vec{F}_1$. If not, explain why not.

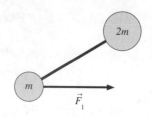

26. Forces $\vec{F}_1$ and $\vec{F}_2$ have the same magnitude and are applied to the corners of a square plate. Is there a *single* force $\vec{F}_3$ that, if applied to the appropriate point on the plate, will cause the plate to be in total equilibrium? If so, draw it, making sure it has the right position, orientation, and length. If not, explain why not.

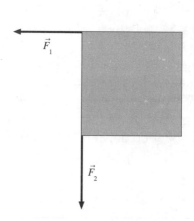

27. The steel girder of a bridge is supported by four posts. A truck is driving across the bridge and is at the position shown.

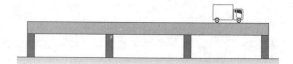

   a. On the figure, draw force arrows to show all forces acting on the steel girder.

   b. Is the net torque about the left support post positive, negative, or zero? Explain.

# 13.7 Rotational Energy

28. The figure shows four identical particles moving in circles. Rank in order, from largest to smallest, their rotational kinetic energies $K_1$ to $K_4$.

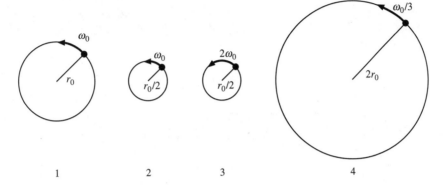

Order:

Explanation:

29. If the angular velocity $\omega$ is held constant, by what *factor* must $R$ change to double the rotational kinetic energy?

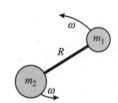

# 13.8 Rolling Motion

30. A wheel is rolling along a horizontal surface with the center-of-mass velocity shown. Draw the velocity vector $\vec{v}$ at points 1 to 4 on the rim of the wheel.

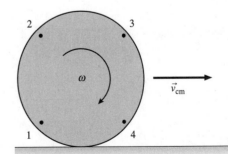

31. A wheel is rolling along a horizontal surface with the center-of-mass velocity shown. Draw the velocity vector $\vec{v}$ at points 1 to 3 on the wheel.

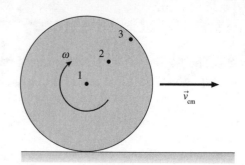

32. If a solid disk and a circular hoop of the same mass and radius are released from rest at the top of a ramp and allowed to roll to the bottom, the disk will get to the bottom first. *Without referring to equations*, explain why this is so.

## 13.9  The Vector Description of Rotational Motion

## 13.10  Angular Momentum of a Rigid Body

33. For each vector pair $\vec{A}$ and $\vec{B}$ shown below, determine if $\vec{A} \times \vec{B}$ points into the page, out of the page, or is zero.

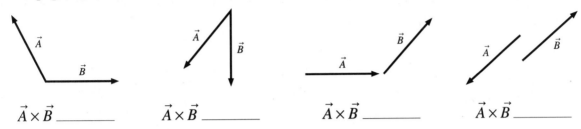

$\vec{A} \times \vec{B}$ _____     $\vec{A} \times \vec{B}$ _____     $\vec{A} \times \vec{B}$ _____     $\vec{A} \times \vec{B}$ _____

34. Each figure below shows $\vec{A}$ and $\vec{A} \times \vec{B}$. Determine if $\vec{B}$ is in the plane of the page or perpendicular to the page. If $\vec{B}$ is in the plane of the page, draw it. If $\vec{B}$ is perpendicular to the page, state whether $\vec{B}$ points into the page or out of the page.

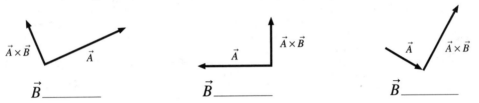

$\vec{B}$ _____          $\vec{B}$ _____          $\vec{B}$ _____

35. Draw the angular velocity vector on each of the rotating wheels.

a.                    b.                    c.

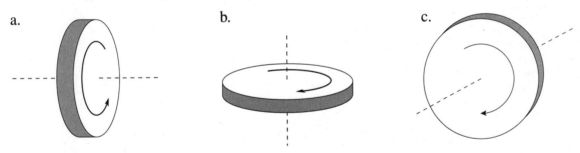

36. The figures below show a force acting on a particle. For each, draw the torque vector for the torque about the origin.
   - Place the tail of the torque vector at the origin.
   - Draw the vector large and straight (use a ruler!) so that its direction is clear. Use dotted lines from the tip of the vector to the axes to show the plane in which the vector lies.

a.                    b.                    c.

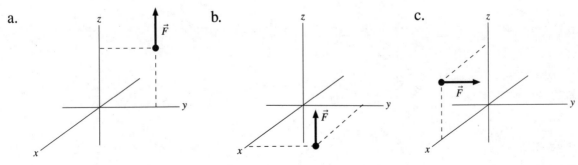

37. The figures below show a particle with velocity $\vec{v}$. For each, draw the angular momentum vector $\vec{L}$ for the angular momentum relative to the origin. Place the tail of the angular momentum vector at the origin.

a.
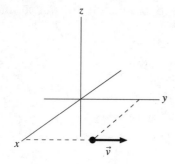

b.
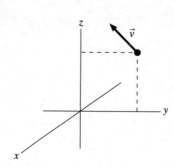

c.
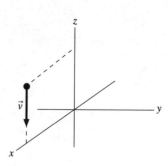

38. Rank in order, from largest to smallest, the angular momenta $L_1$ to $L_4$.

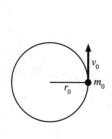

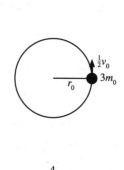

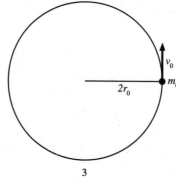

1                 2                 3                 4

Order:

Explanation:

39. Is the angular momentum of disk 2 larger than, smaller than, or equal to the angular momentum of disk 1? Explain.

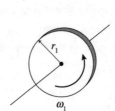

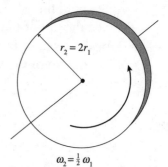

# 14 Oscillations

## 14.1 Simple Harmonic Motion

1. Give three examples of *oscillatory* motion. (Note that circular motion is not the same as oscillatory motion.)

2. On the axes below, sketch three cycles of the displacement-versus-time graph for:

   a. A particle undergoing symmetric periodic motion that is *not* SHM.

   b. A particle undergoing asymmetric periodic motion.

   c. A particle undergoing simple harmonic motion.

3. Consider the particle whose motion is represented by the *x*-versus-*t* graph below.

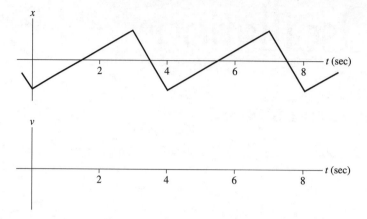

   a. Is this periodic motion? _____     b. Is this motion SHM? _____

   c. What is the period? _____     d. What is the frequency? _____

   e. You learned in Chapter 2 to relate velocity graphs to position graphs. Use that knowledge to draw the particle's velocity-versus-time graph on the axes provided.

4. Shown below is the velocity-versus-time graph of a particle.

   a. What is the period of the motion? _____

   b. Draw the particle's position-versus-time graph, starting from $x = 0$ at $t = 0$ s.

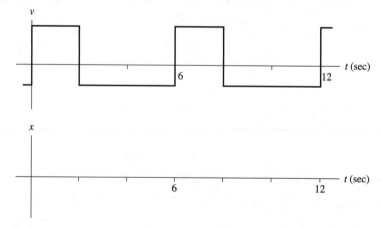

5. The graph on the next page is the position-versus-time graph of an oscillating particle. It is constructed of *parabolic* segments that are joined at $x = 0$.

   a. Is this simple harmonic motion? Why or why not?

   b. Draw the corresponding velocity-versus-time graph.
      Hint: What is the derivative of a parabolic function?

   c. Draw the corresponding acceleration-versus-time graph.

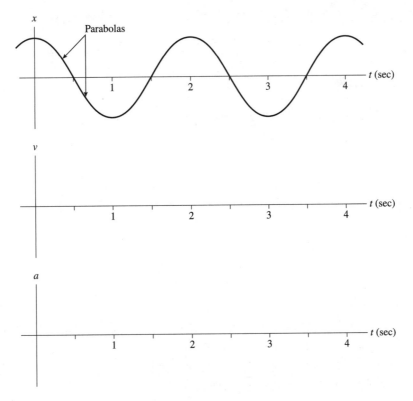

d. At what times is the position a maximum? _____

   At those times, is the velocity a maximum, a minimum, or zero? _____

   At those times, is the acceleration a maximum, a minimum, or zero? _____

e. At what times is the position a minimum (most negative)? _____

   At those times, is the velocity a maximum, a minimum, or zero? _____

   At those times, is the acceleration a maximum, a minimum, or zero? _____

f. At what times is the velocity a maximum? _____

   At those times, where is the particle? _____

g. Can you find a simple relationship between the *sign* of the position and the *sign* of the acceleration at the same instant of time? If so, what is it?

6. The figure shows the position-versus-time graph of a particle in SHM.

   a. At what time or times is the particle moving to the right at maximum speed?

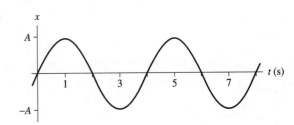

   b. At what time or times is the particle moving to the left at maximum speed?

   c. At what time or times is the particle instantaneously at rest?

# 14.2 Simple Harmonic Motion and Circular Motion

7. A particle goes around a circle 5 times at constant speed, taking a total of 2.5 seconds.

   a. Through what angle *in degrees* has the particle moved? _____

   b. Through what angle *in radians* has the particle moved? _____

   c. What is the particle's frequency $f$?

   d. Use your answer to part b to determine the particle's angular frequency $\omega$.

   e. Does $\omega$ (in rad/s) $= 2\pi f$ (in Hz)? _____

8. A particle moves counterclockwise around a circle at constant speed. For each of the phase constants given below:
   • Show with a dot *on the circle* the particle's starting position.
   • Sketch two cycles of the particle's $x$-versus-$t$ graph.

a.

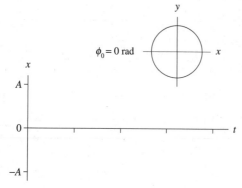

b.

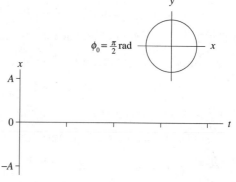

c.

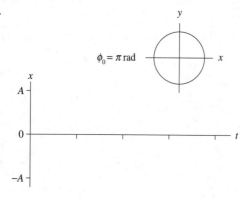

d.

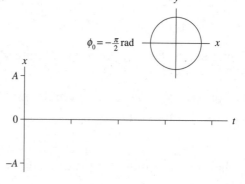

9. a. On the top set of axes below, sketch two cycles of the *x*-versus-*t* graphs for a particle in simple harmonic motion with phase constants i) $\phi_0 = \pi/2$ rad and ii) $\phi_0 = -\pi/2$ rad.

   b. Use the bottom set of axes to sketch velocity-versus-time graphs for the particles. Make sure each velocity graph aligns vertically with the correct points on the *x*-versus-*t* graph.

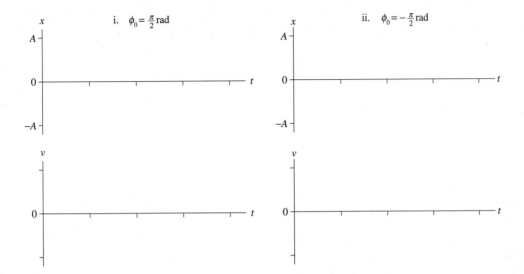

10. The graph below represents a particle in simple harmonic motion.

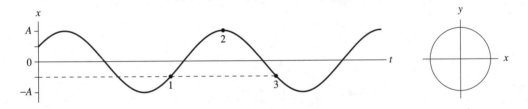

   a. What is the phase constant $\phi_0$? Explain how you determined it.

   b. What is the phase of the particle at each of the three numbered points on the graph?

   Phase at 1: _____   Phase at 2: _____   Phase at 3: _____

   c. Place dots on the circle above to show the position of a circular-motion particle at the times corresponding to points 1, 2, and 3. Label each dot with the appropriate number.

11. The graph shows the *velocity* versus time for a particle in simple harmonic motion.

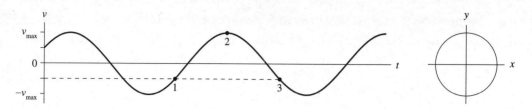

a. What is the phase constant $\phi_0$? Explain how you determined it.

b. What is the phase of the particle at each of the three labeled points on the graph?

Phase at 1: _____ Phase at 2: _____ Phase at 3: _____

c. Place dots on the circle to show the position of a circular-motion particle at the times corresponding to points 1, 2, and 3. Label each dot with the appropriate number.

## 14.3  Energy in Simple Harmonic Motion

12. The figure shows the potential-energy diagram and the total energy line of a particle oscillating on a spring.

    a. What is the spring's equilibrium length?

    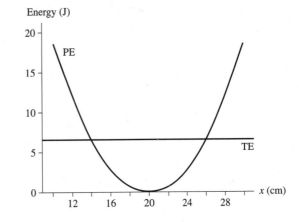

    b. Where are the turning points of the motion? Explain how you identify them.

    c. What is the particle's maximum kinetic energy?

    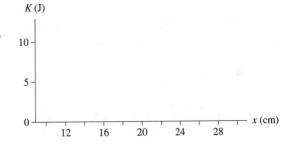

    d. Draw a graph of the particle's kinetic energy as a function of position.

    e. What will be the turning points if the particle's total energy is doubled?

13. A block oscillating on a spring has an amplitude of 20 cm. What will be the block's amplitude if its total energy is doubled? Explain.

14. A block oscillating on a spring has a maximum speed of 20 cm/s. What will be the block's maximum speed if its total energy is doubled? Explain.

15. The figure shows the potential energy diagram of a particle.

    a. Is the particle's motion periodic? How can you tell?

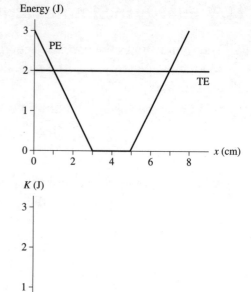

    b. Is the particle's motion simple harmonic motion? How can you tell?

    c. What is the amplitude of the motion?

    d. Draw a graph of the particle's kinetic energy as a function of position.

16. Equation 14.25 in the textbook states that $\frac{1}{2}kA^2 = \frac{1}{2}mv_{max}^2$. What does this mean? Write a couple of sentences explaining how to interpret this equation.

## 14.4 The Dynamics of Simple Harmonic Motion

## 14.5 Vertical Oscillations

17. A block oscillating on a spring has period $T = 2$ s.

    a. What is the period if the block's mass is doubled? Explain.
       **Note:** You do not know values for either $m$ or $k$. Do *not* assume any particular values for them. The required analysis involves thinking about ratios.

    b. What is the period if the value of the spring constant is quadrupled?

    c. What is the period if the oscillation amplitude is doubled while $m$ and $k$ are unchanged?

18. For graphs a and b, determine:
    • The angular frequency $\omega$.
    • The oscillation amplitude $A$.
    • The phase constant $\phi_0$.
    **Note:** Graphs a and b are independent. Graph b is *not* the velocity graph of a.

a.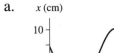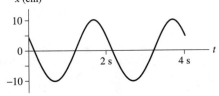

b.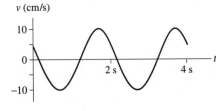

$\omega =$ _____

$A =$ _____

$\phi_0 =$ _____

$\omega =$ _____

$A =$ _____

$\phi_0 =$ _____

19. The graph on the right is the position-versus-time graph for a simple harmonic oscillator.

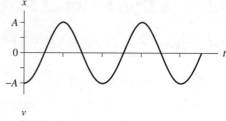

a. Draw the v-versus-t and a-versus-t graphs.

b. When x is greater than zero, is a ever greater than zero? If so, at which points in the cycle?

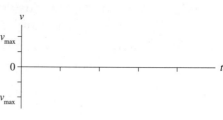

c. When x is less than zero, is a ever less than zero? If so, at which points in the cycle?

d. Can you make a general conclusion about the relationship between the sign of x and the sign of a?

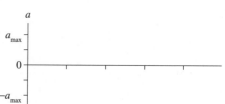

e. When x is greater than zero, is v ever greater than zero? If so, how is the oscillator moving at those times?

20. For the oscillation shown on the left below:

a. What is the phase constant $\phi_0$? _____

b. Draw the corresponding v-versus-t graph on the axes below the x-versus-t graph.

c. On the axes on the right, sketch two cycles of the x-versus-t and the v-versus-t graphs if the value of $\phi_0$ found in part a is replaced by its negative, $-\phi_0$.

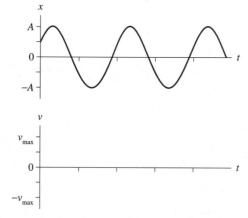
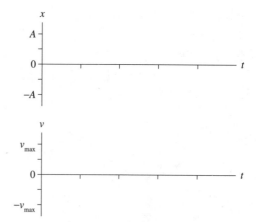

d. Describe *physically* what is the same and what is different about the initial conditions for two oscillators having "equal but opposite" phase constants $\phi_0$ and $-\phi_0$.

21. The top graph shows the position versus time for a
    mass oscillating on a spring. On the axes below, sketch the
    position-versus-time graph for this block for the following
    situations:

    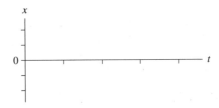

    **Note:** The changes described in each part refer back to the
    original oscillation, not to the oscillation of the previous
    part of the question. Assume that all other parameters
    remain constant. Use the same horizontal and vertical
    scales as the original oscillation graph.

    a. The amplitude and the frequency are doubled.

    b. The amplitude is halved and the mass is quadrupled.

    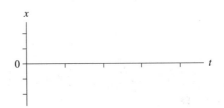

    c. The phase constant is increased by $\pi/2$ rad.

    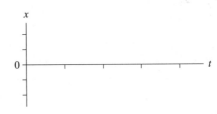

    d. The maximum speed is doubled while the amplitude
       remains constant.

## 14.6 The Pendulum

22. A pendulum on planet X, where the value of $g$ is unknown, oscillates with a period of 2 seconds. What is the period of this pendulum if:

   a. Its mass is doubled?
      **Note:** You do not know the values of $m$, $L$, or $g$, so do not assume any specific values.

   b. Its length is doubled?

   c. Its oscillation amplitude is doubled?

23. The graph shows the displacement $s$ versus time for an oscillating pendulum.

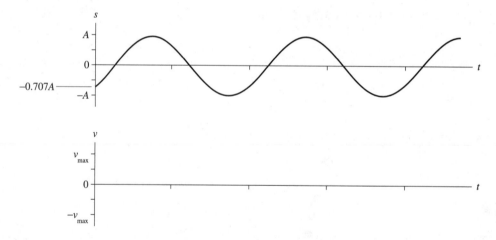

   a. Draw the pendulum's velocity-versus-time graph.
   b. What is the value of the phase constant $\phi_0$?

   c. In the space at the right, draw a *picture* of the pendulum that shows (and labels!)
      • The extremes of its motion.
      • Its position at $t = 0$ s.
      • Its direction of motion (using an arrow) at $t = 0$ s.

# 14.7 Damped Oscillations

24. If the damping constant $b$ of an oscillator is increased,

    a. Is the medium more resistive or less resistive?    _____

    b. Do the oscillations damp out more quickly or less quickly?    _____

    c. Is the time constant $\tau$ increased or decreased?    _____

25. A block on a spring oscillates horizontally on a table with friction. Draw and label force vectors on the block to show all *horizontal* forces on the block.

    a. The mass is to the right of the equilibrium point and moving away from it.

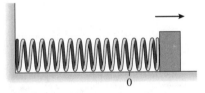

    b. The mass is to the right of the equilibrium point and approaching it.

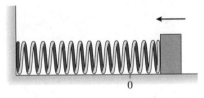

26. The figure below shows the envelope of the oscillations of a damped oscillator. On the same axes, draw the envelope of oscillations if

    a. The time constant is doubled.

    b. The time constant is halved.

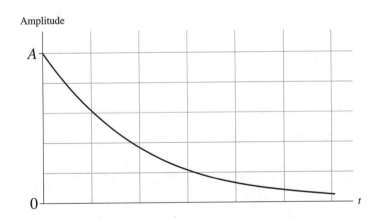

27. a. Describe the difference between $\tau$ and $T$. Don't just *name* them; say what is different about the physical concepts that they represent.

    b. Describe the difference between $\tau$ and $t_{1/2}$.

## 14.8 Driven Oscillations and Resonance

28. What is the difference between the driving frequency and the natural frequency of an oscillator?

29. A car drives along a bumpy road on which the bumps are equally spaced. At a speed of 20 mph, the frequency of hitting bumps is equal to the natural frequency of the car bouncing on its springs.

    a. Draw a graph of the car's vertical bouncing amplitude as a function of its speed if the car has new shock absorbers (large damping coefficient).

    b. Draw a graph of the car's vertical bouncing amplitude as a function of its speed if the car has worn out shock absorbers (small damping coefficient).

    Draw both graphs on the same axes, and label them as to which is which.

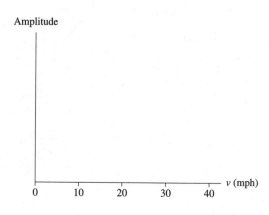

# 15 Fluids and Elasticity

## 15.1 Fluids

1. An object has density $\rho$.

  a. Suppose each of the object's three dimensions is increased by a factor of 2 without changing the material of which the object is made. Will the density change? If so, by what factor? Explain.

  b. Suppose each of the object's three dimensions is increased by a factor of 2 without changing the object's mass. Will the density change? If so, by what factor? Explain.

2. Air enclosed in a cylinder has density $\rho = 1.4 \ \text{kg/m}^3$.

  a. What will be the density of the air if the length of the cylinder is doubled while the radius is unchanged?

  b. What will be the density of the air if the radius of the cylinder is halved while the length is unchanged?

3. Air enclosed in a sphere has density $\rho = 1.4 \ \text{kg/m}^3$. What will the density be if the radius of the sphere is halved?

## 15.2 Pressure

## 15.3 Measuring and Using Pressure

4. When you stand on a bathroom scale, it reads 700 N. Suppose a giant vacuum cleaner sucks half the air out of the room, reducing the pressure to 0.5 atm. Would the scale reading increase, decrease, or stay the same? Explain.

5. Rank in order, from largest to smallest, the pressures at A, B, and C.

   Order:

   Explanation:

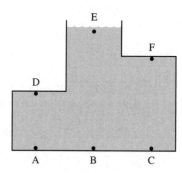

6. Refer to the figure in Exercise 5. Rank in order, from largest to smallest, the pressures at D, E, and F.

   Order:

   Explanation:

7. The gauge pressure at the bottom of a cylinder of liquid is $p_g = 0.4$ atm. The liquid is poured into another cylinder with twice the radius of the first cylinder. What is the gauge pressure at the bottom of the second cylinder?

8. Cylinders A and B contain liquids. The pressure $p_A$ at the bottom of A is higher than the pressure $p_B$ at the bottom of B. Is the ratio $p_A/p_B$ of the absolute pressures larger than, smaller than, or equal to the ratio of the gauge pressures? Explain.

9. A and B are rectangular tanks full of water. They have equal depths, equal thicknesses (the dimension into the page), but different widths.

    a. Compare the forces the water exerts on the bottoms of the tanks. Is $F_A$ larger than, smaller than, or equal to $F_B$? Explain.

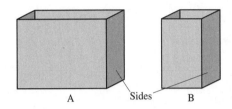
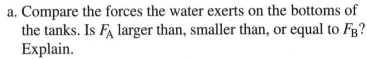

    b. Compare the forces the water exerts on the sides of the tanks. Is $F_A$ larger than, smaller than, or equal to $F_B$? Explain.

10. Water expands when heated. Suppose a beaker of water is heated from 10°C to 90°C. Does the pressure at the bottom of the beaker increase, decrease, or stay the same? Explain.

11. Is $p_A$ larger than, smaller than, or equal to $p_B$? Explain.

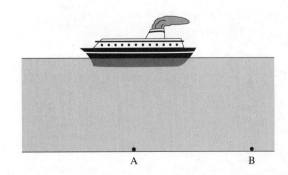

12. The container shown holds a mixture of oil and water. To begin, the container is shaken vigorously to mix the oil into the water by breaking it into very tiny droplets. This is what happens when you shake a jar of salad dressing. Eventually, the oil separates and rises to the top. Oil and water are *immiscible*, meaning that the total volume is the same whether they are mixed or separated. The pressure at the bottom of the container after the oil has separated is *not* the same as the initial pressure when the oil and water are mixed, although it may take some careful thought to understand why.

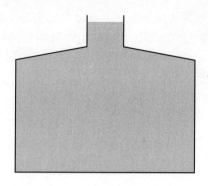

Is the final pressure at the bottom higher or lower than the initial pressure? Explain.

13. At sea level, the height of the mercury column in a sealed glass tube is 380 mm. What can you say about the contents of the space above the mercury? Be as specific as you can.

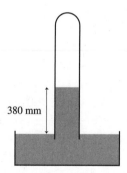

380 mm

# 15.4 Buoyancy

14. Rank in order, from largest to smallest, the densities of A, B, and C.

    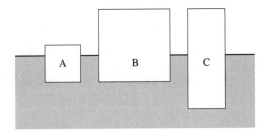

    Order:

    Explanation:

15. A, B, and C have the same volume. Rank in order, from largest to smallest, the sizes of the buoyant forces $F_A$, $F_B$, and $F_C$ on A, B, and C.

    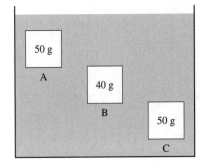

    Order:

    Explanation:

16. Refer to the figure of Exercise 15. Now A, B, and C have the same density. Rank in order, from largest to smallest, the sizes of the buoyant forces on A, B, and C.

    Order:

    Explanation:

17. Suppose you stand on a bathroom scale that is on the bottom of a swimming pool. The water comes up to your waist. Does the scale read more than, less than, or the same as your true weight? Explain.

18. Ships A and B have the same height and the same mass. Their cross section profiles are shown in the figure. Does one ship ride higher in the water (more height above the water line) than the other? If so, which one? Explain.

A

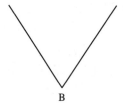
B

# 15.5  Fluid Dynamics

19. Gas flows through a pipe. You can't see into the pipe to know how the inner diameter changes. Rank in order, from largest to smallest, the gas speeds $v_1$ to $v_3$ at points 1, 2, and 3.

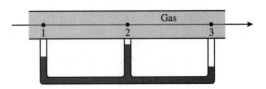

    Order:

    Explanation:

20. Liquid flows through a pipe. You can't see into the pipe to know how the inner diameter changes. Rank in order, from largest to smallest, the flow speeds $v_1$ to $v_3$ at points 1, 2, and 3.

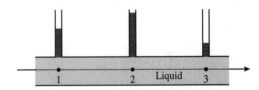

    Order:

    Explanation:

21. Liquid flows through this pipe. This is an overhead view.

    a. Rank in order, from largest to smallest, the flow speeds $v_1$ to $v_4$ at points 1 to 4.

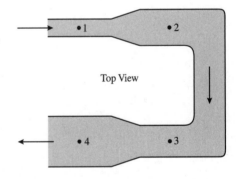

       Order:

       Explanation:

    b. Rank in order, from largest to smallest, the pressures $p_1$ to $p_4$ at points 1 to 4.

       Order:

       Explanation:

22. Wind blows over a house. A window on the ground floor is open. Is there an air flow through the house? If so, does the air flow in the window and out the chimney, or in the chimney and out the window? Explain.

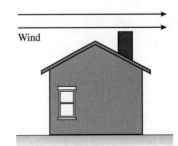

# 15.6 Elasticity

23. A force stretches a wire by 1 mm.

   a. A second wire of the same material has the same cross section and twice the length. How far will it be stretched by the same force? Explain.

   b. A third wire of the same material has the same length and twice the diameter as the first. How far will it be stretched by the same force? Explain.

24. A 2000 N force stretches a wire by 1 mm.

   a. A second wire of the same material is twice as long and has twice the diameter. How much force is needed to stretch it by 1 mm? Explain.

   b. A third wire is twice as long as the first and has the same diameter. How far is it stretched by a 4000 N force?

25. A wire is stretched right to the breaking point by a 5000 N force. A longer wire made of the same material has the same diameter. Is the force that will stretch it right to the breaking point larger than, smaller than, or equal to 5000 N? Explain.

26. Sphere A is compressed by 1% at an ocean depth of 4000 m. Sphere B is compressed by 1% at an ocean depth of 5000 m. Which has the larger bulk modulus? Explain.

# 16 A Macroscopic Description of Matter

## 16.1 Solids, Liquids, and Gases

1. A $1 \times 10^{-3}$ m$^3$ chunk of material has a mass of 3 kg.

   a. What is the material's density?

   b. Would a $2 \times 10^{-3}$ m$^3$ chunk of the same material have the same mass? Explain.

   c. Would a $2 \times 10^{-3}$ m$^3$ chunk of the same material have the same density? Explain.

2. You are given an irregularly-shaped chunk of material and asked to find its density. List the *specific* steps that you would follow to do so.

3. Object 1 has an irregular shape. Its density is 4000 kg/m$^3$.

   a. Object 2 has the same shape and dimensions as object 1, but it is twice as massive. What is the density of object 2?

   b. Object 3 has the same mass and the same *shape* as object 1, but its size in all three dimensions is twice that of object 1. What is the density of object 3?

## 16.2 Atoms and Moles

4. You have 100 g of aluminum and 100 g of lead.

   a. Which has the greater volume? Explain.

   b. Which contains a larger number of moles? Explain.

   c. Which contains more atoms? Explain.

5. A cylinder contains 2 g of oxygen gas. A piston is used to compress the gas. After the gas has been compressed:

   a. Has the mass of the gas increased, decreased, or not changed? Explain.

   b. Has the density of the gas increased, decreased, or not changed? Explain.

   c. Have the number of moles of gas increased, decreased, or not changed? Explain.

   d. Has the number density of the gas increased, decreased, or not changed? Explain.

# 16.3 Temperature

6. Rank in order, from highest to lowest, the temperatures $T_1 = 0$ K, $T_2 = 0°C$, and $T_3 = 0°F$.

7. "Room temperature" is often considered to be 68°F. What is room temperature in °C and in K?

8. a. The gas pressure inside a sealed, rigid container is 1 atm at 100K. What is the pressure at 200K?

   b. The gas pressure inside a sealed, rigid container is 1 atm at 100°C. What is the pressure at 200°C?

## 16.4 Phase Changes

9. On the phase diagram:

   a. Draw *and label* a line to show a process in which the substance boils at constant pressure. Be sure to include an arrowhead on your line to show the direction of the process.

   b. Draw *and label* a line to show a process in which the substance freezes at constant temperature.

   c. Draw *and label* a line to show a process in which the substance sublimates at constant pressure.

   d. Draw a small circle around the critical point.

   e. Draw a small box around the triple point.

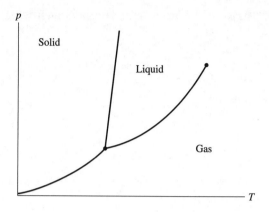

10. The figure shows the phase diagram of water. Answer the following questions by referring to the phase diagram and, perhaps, drawing lines on the phase diagram.

    a. What happens to the boiling-point temperature of water as you go to higher and higher elevations in the mountains?

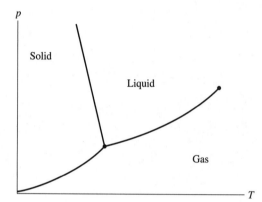

    b. Suppose you place a beaker of liquid water at 20°C in a vacuum chamber and then steadily reduce the pressure. What, if anything, happens to the water?

    c. Is ice less dense or more dense than liquid water?

# 16.5 Ideal Gases

11. It is well known that you can trap liquid in a drinking straw by placing the tip of your finger over the top while the straw is in the liquid, then lifting it out. The liquid runs out when you release your finger.

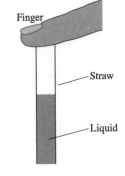

a. What is the *net* force on the cylinder of trapped liquid?

b. Draw a free-body diagram for the trapped liquid. Label each vector.

c. Is the gas pressure inside the straw, between the liquid and your finger, greater than, less than, or equal to atmospheric pressure? Explain.

d. If your answer to part c was "greater" or "less," how did the pressure change from the atmospheric pressure that was present when you placed your finger over the top of the straw?

12. A gas is in a sealed container. By what factor does the gas pressure change if:

a. The volume is doubled and the temperature is tripled?

b. The volume is halved and the temperature is tripled?

13. A gas is in a sealed container. By what factor does the gas temperature change if:
    a. The volume is doubled and the pressure is tripled?

    b. The volume is halved and the pressure is tripled?

14. The gas inside in a cylinder is heated, causing a piston in the cylinder to move outward. The heating causes the temperature to double and the length of the cylinder to triple. By what factor does the gas pressure change?

15. A gas is in a sealed container. The gas pressure is tripled and the temperature is doubled.
    a. What happens to the number of moles of gas in the container?

    b. What happens to the number density of the gas in the container?

## 16.6 Ideal Gas Processes

16. The graphs below show the initial state of a gas. Draw a $pV$ diagram showing the following processes:

a. An isochoric process that doubles the pressure.

b. An isobaric process that doubles the temperature.

c. An isothermal process that halves the volume.

 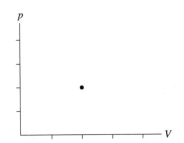 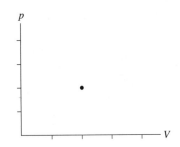

17. Interpret the $pV$ diagrams shown below by

a. Naming the process.

b. Stating the *factors* by which $p$, $V$, and $T$ change. (A fixed quantity changes by a factor of 1.)

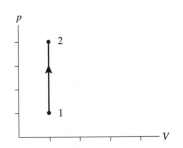 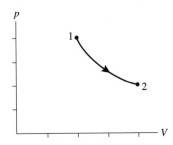 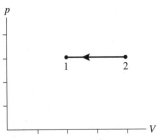

Process _____     Process _____     Process _____

$p$ changes by _____     $p$ changes by _____     $p$ changes by _____

$V$ changes by _____     $V$ changes by _____     $V$ changes by _____

$T$ changes by _____     $T$ changes by _____     $T$ changes by _____

18. Starting from the initial state shown, draw a $pV$ diagram for the three-step process:

i.   An isochoric process that halves the temperature,

ii.  An isothermal process that halves the pressure, then

iii. An isobaric process that doubles the volume.

Label each of the stages on your diagram.

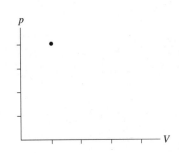

# 17 Work, Heat, and the First Law of Thermodynamics

## 17.1 It's All About Energy

## 17.2 Work in Ideal-Gas Processes

1. How much work is done on the gas in each of the following processes?

a.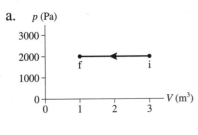

W = _____

b.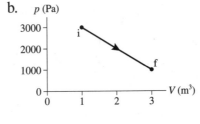

W = _____

c.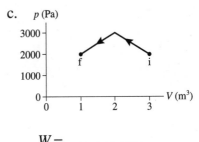

W = _____

2. The figure below shows a process in which a gas is compressed from 300 cm³ to 100 cm³.

a. Use the middle set of axes to draw the pV diagram of a process that starts from initial state i, compresses the gas to 100 cm³, and does the same amount of work on the gas as the process on the left.

b. Is there an isochoric process that does the same amount of work on the gas as the process on the left? If so, show it on the axes on the right. If not, use the blank space of the axes to explain why.

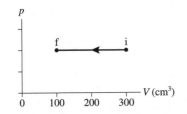 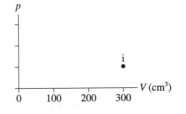 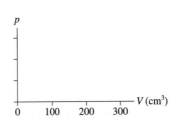

3. The figure shows a process in which work is done to compress a gas.

a. Draw and label a process A that starts and ends at the same points but does *more* work on the gas.

b. Draw and label a process B that starts and ends at the same points but does *less* work on the gas.

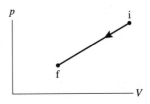

## 17.3 Heat

## 17.4 The First Law of Thermodynamics

4. When the space shuttle returns to earth, its surfaces get very hot as it passes through the atmosphere at high speed.

   a. Has the space shuttle been heated? If so, what was the source of the heat? If not, why is it hot?

   b. Energy must be conserved. What happens to the space shuttle's initial kinetic energy?

5. Cold water is poured into a hot metal container.

   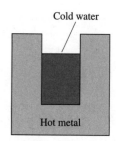

   a. What physical quantities can you *measure* that tell you that the metal and water are somehow changing?

   b. What is the condition for equilibrium, after which no additional changes take place?

   c. Use the concept of energy to describe how the metal and the water interact.

   d. Is your description in part c something that you can *observe* happening? Or is it an *inference* based on the measurements you specified in part a?

6. Do each of the following describe a property of a system, an interaction of a system with its environment, or both? Explain.

 a. Temperature:

 b. Heat:

 c. Thermal energy:

7. a. For each of the following processes:

 i.   Is the value of the work $W$, the heat $Q$, and the change of thermal energy $\Delta E_{th}$ positive (+), negative (−) or zero (0)?

 ii.  Does the temperature increase (+), decrease (−), or not change (0)?

| | $W$ | $Q$ | $\Delta E_{th}$ | $\Delta T$ |
|---|---|---|---|---|
| You hit a nail with a hammer. | ____ | ____ | ____ | ____ |
| You hold a nail over a Bunsen burner. | ____ | ____ | ____ | ____ |
| You compress the air in a bicycle pump by pushing down on the handle very rapidly. | ____ | ____ | ____ | ____ |
| You turn on a flame under a cylinder of gas, and the gas undergoes an isothermal expansion. | ____ | ____ | ____ | ____ |
| A flame turns liquid water into steam. | ____ | ____ | ____ | ____ |
| High-pressure steam spins a turbine. | ____ | ____ | ____ | ____ |
| Steam contacts a cold surface and condenses. | ____ | ____ | ____ | ____ |
| A moving crate slides to a halt on a rough surface. | ____ | ____ | ____ | ____ |
| High-pressure gas in a cylinder pushes a piston outward very rapidly. | ____ | ____ | ____ | ____ |

 b. Are each of your responses consistent with the first law of thermodynamics? If not, which ones are not?

8. The text says that the first law of thermodynamics is simply a general statement of the idea of conservation of energy. What does this mean? How does the first law embody the idea of energy conservation?

9. Consider an ideal-gas process that increases the volume of the gas in a cylinder without changing its pressure.

   a. Show the process on a $pV$ diagram.

   b. Show the process on a first-law bar chart.

   c. What kind of process is this? _____

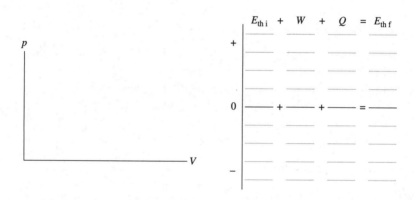

10. Consider an ideal-gas process that increases the pressure of the gas in a cylinder without changing its temperature.

   a. Show the process on a $pV$ diagram.

   b. Show the process on a first-law bar chart.

   c. What kind of process is this? _____

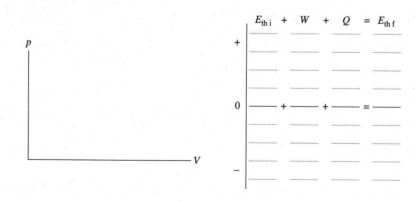

# 17.5  Thermal Properties of Matter

# 17.6  Calorimetry

11. A beaker of water at 80.0°C is placed in the center of a large, well-insulated room whose air temperature is 20.0°C. Is the final temperature of the water:

    i.   20.0°C.
    ii.  Slightly above 20.0°C.
    iii. 50.0°C.

    iv.  Slightly less than 80.0°C.
    v.   80.0°C.

    Explain.

12. Materials A and B have equal densities, but A has a larger specific heat than B. You have 100 g cubes of each material.

    a. Cube A, initially at 0°C, is placed in good thermal contact with cube B, initially at 200°C. The cubes are inside a well-insulated container where they don't interact with their surroundings. Is their final temperature greater than, less than, or equal to 100°C? Explain.

    b. Cube A and cube B are both heated to 200°C, then placed on a table in room-temperature air. Which one cools down more quickly? Explain.

13. 100 g of ice at 0°C and 100 g of steam at 100°C interact thermally in a well-insulated container. Is the final state of the system

    i.   An ice-water mixture at 0°C?
    ii.  Water at a temperature between 0°C and 50°C?
    iii. Water at 50°C?
    iv.  Water at a temperature between 50°C and 100°C?
    v.   A water-steam mixture at 100°C?

    Explain.

## 17.7 The Specific Heats of Gases

14. Two containers hold equal masses of nitrogen gas at equal temperatures. You supply 10 J of heat to container A while not allowing its volume to change, and you supply 10 J of heat to container B while not allowing its pressure to change. Afterward, is temperature $T_A$ greater than, less than, or equal to $T_B$? Explain.

15. You need to raise the temperature of a gas by 10°C. To use the least amount of heat energy, should you heat the gas at constant pressure or at constant volume? Explain.

16. Describe *why* the molar specific heat at constant pressure is larger than the molar specific heat at constant volume.

17. The figure shows an adiabatic process.
    a. Is the final temperature higher than, lower than, or equal to the initial temperature?

    b. Draw *and label* the $T_i$ and $T_f$ isotherms on the figure.
    c. Is the work done on the gas positive or negative? Explain.

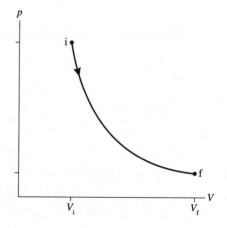

    d. Show *on the figure* how you would determine the amount of work done.
    e. Is any heat energy added to or removed from the system in this process? Explain.

    f. *Why* does the gas temperature change?

# 18 The Micro/Macro Connection

## 18.1 Molecular Collisions

1. a. Solids and liquids resist being compressed. They are not totally incompressible, but it takes large forces to compress them even slightly. If it is true that matter consists of atoms, what can you infer about the microscopic nature of solids and liquids from their incompressibility?

   b. Solids, and to a lesser extent liquids, also resist being pulled apart. You can break a metal or glass rod by pulling the ends in opposite directions, but it takes a large force to do so. What can you infer from this observation about the properties of atoms?

2. a. Gases, in contrast with solids and liquids, are very compressible. What can you infer from this observation about the microscopic nature of gases?

   b. The density of air at STP is about $\frac{1}{1000}$ the density of water. How does the average distance between air molecules compare to the average distance between water molecules? Explain.

3. a. Can you think of any everyday experiences or observations that would suggest that the molecules of a gas are in constant, *random* motion? (Note: The existence of "wind" is *not* such an observation. Wind implies that the gas as a whole can move, but it doesn't tell you anything about the motions of the individual molecules in the gas.)

   b. If the molecules are moving randomly and colliding with each other, do you expect there to be a very wide range of molecular speeds? Or do you expect that most of the molecules will have very similar speeds? Explain.

4. The mean free path of molecules in a gas is 200 nm.

   a. What will be the mean free path if the pressure is doubled while all other state variables are held constant?

   b. What will be the mean free path if the absolute temperature is doubled while all other state variables are held constant?

5. Helium has atomic mass number $A = 4$. Neon has $A = 20$ and argon has $A = 40$. Rank in order, from largest to smallest, the mean free paths $\lambda_{He}$, $\lambda_{Ne}$, and $\lambda_{Ar}$ at STP. Explain.

## 18.2  Pressure in a Gas

6. If the pressure of a gas is really due to the *random* collisions of molecules with the walls of the container, why do pressure gauges—even very sensitive ones—give perfectly steady readings? Shouldn't the gauge be continually jiggling and fluctuating? Explain.

7. According to kinetic theory, the pressure of a gas depends on the number density and the rms speed of the gas molecules. Consider a sealed container of gas that is heated at constant volume.

   a. According to the ideal gas law, does the pressure of the gas increase or stay the same? Explain.

   b. Does the number density of the gas increase or stay the same? Explain.

   c. What can you infer from these observations about a relationship between the gas temperature (a macroscopic parameter) and the rms speed of the molecules (a microscopic parameter)?

8. Suppose you could suddenly increase the speed of every molecule in a gas by a factor of 2.

   a. Would the rms speed of the molecules increase by a factor of $(2)^{1/2}$, 2, or $2^2$? Explain.

   b. Would the gas pressure increase by a factor of $(2)^{1/2}$, 2, or $2^2$? Explain.

# 18.3 Temperature

9. If you double the temperature of a gas:

a. Does the average translational kinetic energy per molecule change? If so, by what factor?

b. Does the root-mean-square velocity of the molecules change? If so, by what factor?

10. Lithium vapor, which is produced by heating lithium to the relatively low boiling point of 1340°C, forms a gas of $Li_2$ molecules. Each molecule has a molecular mass of 14 u. The molecules in nitrogen gas ($N_2$) have a molecular mass of 28 u. If the $Li_2$ and $N_2$ gases are at the same temperature, which of the following is true?

i.   $v_{rms}$ of $N_2 = 2.00 \times v_{rms}$ of $Li_2$.
ii.  $v_{rms}$ of $N_2 = 1.41 \times v_{rms}$ of $Li_2$.
iii. $v_{rms}$ of $N_2 = v_{rms}$ of $Li_2$.
iv.  $v_{rms}$ of $N_2 = 0.71 \times v_{rms}$ of $Li_2$.
v.   $v_{rms}$ of $N_2 = 0.50 \times v_{rms}$ of $Li_2$.

Explain.

11. Suppose you could suddenly increase the speed of every molecule in a gas by a factor of 2. Would the temperature of the gas increase by a factor of $(2)^{1/2}$, 2, or $2^2$? Explain.

12. Two gases have the same number density and the same distribution of speeds. The molecules of gas 2 are more massive than the molecules of gas 1.

   a. Do the two gases have the same pressure? If not, which is larger?

   b. Do the two gases have the same temperature? If not, which is larger?

13. a. What is the average translational kinetic energy of a gas at absolute zero?

   b. Can a molecule have negative translational kinetic energy? Explain.

   c. Based on your answers to parts a and b, what is the translational kinetic energy of *every* molecule in the gas?

   d. Would it be physically possible for the thermal energy of a gas to be less than its thermal energy at absolute zero? Explain.

   e. Is it possible to have a temperature less than absolute zero? Explain.

# 18.4  Thermal Energy and Specific Heat

# 18.5  Thermal Interactions and Heat

14. Suppose you could suddenly increase the speed of every molecule in a gas by a factor of 2.

    a. Does the thermal energy of the gas change? If so, by what factor? If not, why not?

    b. Does the molar specific heat change? If so, by what factor? If not, why not?

15. Hot water is poured into a cold container. Give a *microscopic* description of how these two systems interact until they reach thermal equilibrium.

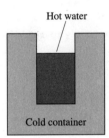

Hot water

Cold container

16. A beaker of cold water is placed over a flame.

    a. What *is* a flame?

Water

Flame

    b. Give a *microscopic* description of how the flame increases the water temperature.

17. The *rapid* compression of a gas by a fast-moving piston increases the gas temperature. For example, you likely have noticed that pumping up a bicycle tire causes the bottom of the pump to get warm. Consider a gas that is rapidly compressed by a piston.

a. Does the thermal energy of the gas increase or stay the same? Explain.

b. Is there a transfer of heat energy to the gas? Explain.

c. Is work done on the gas? Explain.

d. Give a *microscopic* description of why the gas temperature increases as the piston moves in.

18. A container with 0.1 mol of helium ($A = 4$) and a container with 0.2 mol of argon ($A = 40$) are placed in thermal contact with each other. The helium has an initial temperature of 200°C and the argon has an initial temperature of 0°C. *After* they have reached thermal equilibrium:

| 0.1 mol He | 0.2 mol Ar |
|---|---|
| 200°C | 0°C |

a. Is $v_{rms}$ of helium greater than, less than, or equal to $v_{rms}$ of argon? Explain.

b. Does the helium have more thermal energy than, less thermal energy than, or the same amount of thermal energy as the argon? Explain.

# 18.6  Irreversible Processes and the Second Law of Thermodynamics

19. Every cubic meter of air contains ≈250,000 J of thermal energy. This is approximately the kinetic energy of a car going 50 mph. Even though it might be difficult to do, could a clever engineer design a car that uses the thermal energy already in the air as "fuel"? Even if only 1% of the thermal energy could be "extracted" from the air, it would take only ≈100 m$^3$ of air—the volume of a typical living room in a house—to get the car up to speed. Is this idea possible? Or does it violate the laws of physics?

20. If you place a jar of perfume in the center of a room and remove the stopper, you will soon be able to smell the perfume throughout the room. If you wait long enough, will all the perfume molecules ever be back in the jar at the same time? Why or why not?

21. Suppose you place an ice cube in a cup of room-temperature water and then seal them in a well-insulated container. No energy can enter or leave the container.

   a. If you open the container an hour later, which do you expect to find: a cup of water, slightly cooler than room temperature, or a large ice cube and some 100°C steam?

   b. Finding a large ice cube and some 100°C steam would not violate the first law of thermodynamics. $W = 0$ J and $Q = 0$ J, because the container is sealed, and $\Delta E_{th} = 0$ J because the increase in thermal energy of the water molecules that have become steam is offset by the decrease in water molecules that have turned to ice. Energy is conserved, yet we never see a process like this. Why not?

# 19 Heat Engines and Refrigerators

## 19.1 Turning Heat into Work

## 19.2 Heat Engines and Refrigerators

1. The figure on the left shows a thermodynamic process in which a gas expands from $100 \text{ cm}^3$ to $300 \text{ cm}^3$. On the right, draw the $pV$ diagram of a process that starts from state i, expands to $300 \text{ cm}^3$, and does the same amount of work as the process on the left.

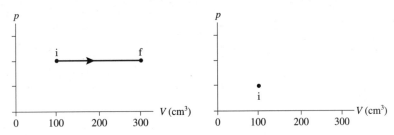

2. For each of these processes, is work done *by* the system ($W < 0$, $W_s > 0$), *on* the system ($W > 0$, $W_s < 0$), or is *no* work done?

a.

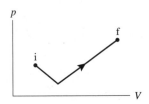

Work is _____

b.

Work is _____

c.

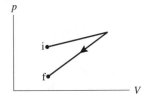

Work is _____

3. Rank in order, from largest to smallest, the thermal efficiencies $\eta_1$ to $\eta_4$ of these heat engines.

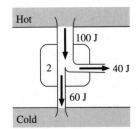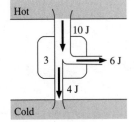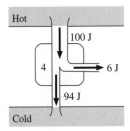

Order:

Explanation:

4. Could you have a heat engine with $\eta > 1$? Explain.

5. For each engine shown,
   a. Supply the missing value.
   b. Determine the thermal efficiency.

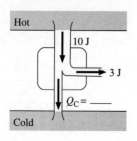

$\eta =$ _____

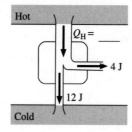

$\eta =$ _____

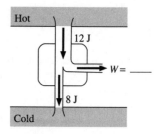

$\eta =$ _____

6. For each refrigerator shown,
   a. Supply the missing value.
   b. Determine the coefficient of performance.

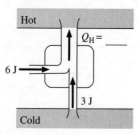

$K =$ _____

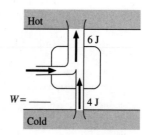

$K =$ _____

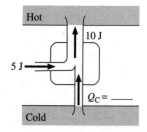

$K =$ _____

7. Does a refrigerator do work in order to cool the interior? Explain.

# 19.3  Ideal-Gas Heat Engines

8. Starting from the point shown, draw a $pV$ diagram for the following processes.

   a. An isobaric process in which work is done *by* the system.

   b. An adiabatic process in which work is done *on* the system.

   c. An isothermal process in which heat is *added to* the system.

   d. An isochoric process in which heat is *removed from* the system.

9. Rank in order, from largest to smallest, the amount of work $W_1$ to $W_4$ done by the gas in each of these cycles.

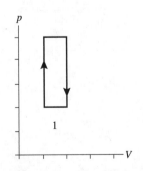

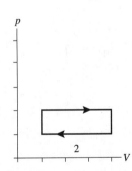

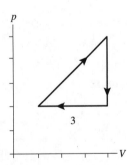

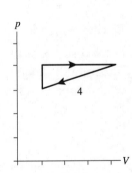

Order:

Explanation:

10. The figure uses a series of pictures to illustrate a thermodynamic cycle.

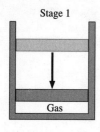

Stage 1

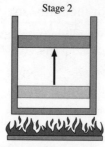

Stage 2

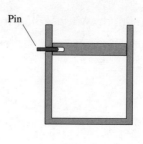

Pin

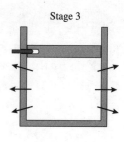

Stage 3

The gas is compressed rapidly from $V_1$ to $V_2$.

The gas is heated at constant temperature until the volume returns to $V_1$.

The flame is turned off and the piston is locked in place.

The gas cools until the initial pressure $p_1$ is restored.

a. Show the cycle as a $pV$ diagram. Label the three stages.

b. What is the energy transformation during each stage of the process? (For example, a stage in which work energy is transformed into heat energy could be represented as $W \rightarrow Q$.)

Stage 1: _____

Stage 2: _____

Stage 3: _____

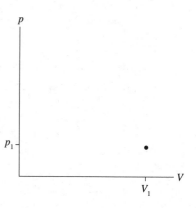

11. The figure shows the $pV$ diagram of a heat engine.

a. During which stages is heat added to the gas? _____

b. During which is heat removed from the gas? _____

c. During which stages is work done on the gas? _____

d. During which is work done by the gas? _____

e. Draw a series of pictures, similar to those of Exercise 10, to illustrate the stages of this cycle. Give a brief description of what happens during each stage.

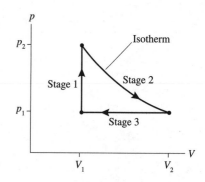

12. A heat engine satisfies $W_{out} = Q_{net}$. Why is there no $\Delta E_{th}$ term in this relationship?

13. Two thermodynamic cycles are shown. Which cycle has a larger thermal efficiency? Explain.

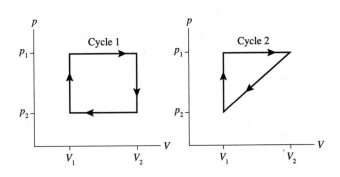

14. Two thermodynamic cycles are shown. Which cycle has a larger thermal efficiency? Explain.

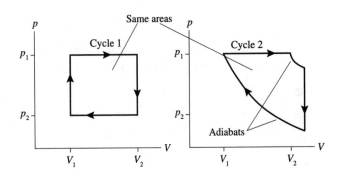

# 19.4 Ideal-Gas Refrigerators

15. a. The figure shows the $pV$ diagram of a heat engine.

During which stages is heat added to the gas? _____

During which is heat removed from the gas? _____

During which stages is work done on the gas? _____

During which is work done by the gas? _____

Circle the answers that complete the sentence:
The temperature of the hot reservoir must be

      i. >      ii. =      iii. <

the temperature  i. $T_2$  ii. $T_3$  iii. $T_4$

Circle the answers that complete the sentence: The temperature of the cold reservoir must be

      i. >      ii. =      iii. <

the temperature  i. $T_2$  ii. $T_1$  iii. $T_4$

b. The figure shows the $pV$ diagram of a refrigerator.

During which stages is heat added to the gas? _____

During which is heat removed from the gas? _____

During which stages is work done on the gas? _____

During which is work done by the gas? _____

Circle the answers that complete the sentence:
The temperature of the hot reservoir must be

      i. >      ii. =      iii. <

the temperature  i. $T_2$  ii. $T_3$  iii. $T_4$

Circle the answers that complete the sentence:
The temperature of the cold reservoir must be

      i. >      ii. =      iii. <

the temperature  i. $T_2$  ii. $T_1$  iii. $T_4$

16. An ideal-gas device operates with the cycle shown. Is it a refrigerator? That is, does it remove heat energy from a cold side and exhaust heat energy to a hot side? Explain.

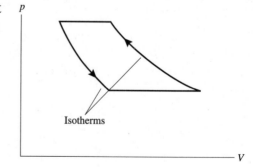

## 19.5  The Limits of Efficiency

## 19.6  The Carnot Cycle

17. Do each of the following represent a possible heat engine or refrigerator? If not, what is wrong?

a.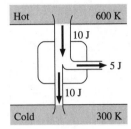

b.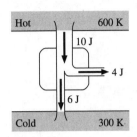

c.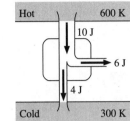

d.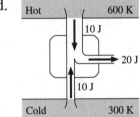

e.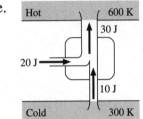

f.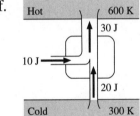

18. Four Carnot engines operate with the hot and cold reservoir temperatures shown in the table.

| Engine | $T_C$ (K) | $T_H$ (K) |
|---|---|---|
| 1 | 300 | 600 |
| 2 | 200 | 400 |
| 3 | 200 | 600 |
| 4 | 300 | 400 |

Rank in order, from largest to smallest, the thermal efficiencies $\eta_1$ to $\eta_4$ of these engines.

Order:

Explanation:

19. It gets pretty hot in your unairconditioned apartment. In browsing the internet, you find a company selling small "room air conditioners." You place it on the floor, plug it in, and—the advertisement says—the air conditioner will lower the room temperature up to 10°F. Should you order one? Explain.

# 20 Traveling Waves

## 20.1 The Wave Model

1. a. In your own words, define what a *transverse wave* is.

   b. Give an example of a wave that, from your own experience, you know is a transverse wave. What observations or evidence tells you this is a transverse wave?

2. a. In your own words, define what a *longitudinal wave* is.

   b. Give an example of a wave that, from your own experience, you know is a longitudinal wave. What observations or evidence tells you this is a longitudinal wave?

3. Three wave pulses travel along the same string. Rank in order, from largest to smallest, their wave speeds $v_1$, $v_2$, and $v_3$.

   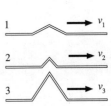

   Order:

   Explanation:

4. The figure shows a wave pulse on a string at time $t = 0$ s. The pulse is traveling to the right at 1000 cm/s, or 1 cm/ms. There is a small bead attached to the string at $x = 9$ cm.

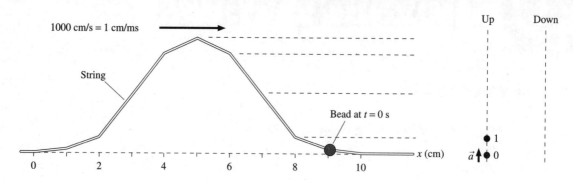

a. Draw a motion diagram of the bead. Show one frame every millisecond, from $t = 0$ ms to $t = 8$ ms. Show upward motion of the bead on the left dotted line and downward motion of the bead on the right line. Show the very top position, at $t = 4$ ms, on *both* lines. The first two frames, at 0 and 1 ms, are already shown and labeled.

b. Draw and label the $\vec{v}$ and $\vec{a}$ vectors on your motion diagram. Write $\vec{a} = \vec{0}$ if appropriate. Recall that velocity vectors go *between* the dots and acceleration vectors, which relate two velocity vectors, go *beside* the dots. The first acceleration vector is already shown.

c. At each instant, the bead feels two tension forces $\vec{T}_R$ and $\vec{T}_L$ along the direction of the string. Draw nine free-body diagrams, from $t = 0$ ms to $t = 8$ ms, showing the two tension forces *and* the net force $\vec{F}_{net}$. Use a different color for $\vec{F}_{net}$. If $\vec{F}_{net} = \vec{0}$, write that beside the figure. The $t = 1$ ms drawing is given as an example.

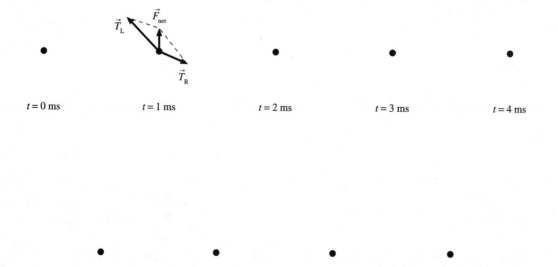

d. Compare your answers to parts b and c. Is Newton's second law obeyed? That is, do $\vec{F}_{net}$ and $\vec{a}$ always point the same direction? If not, why not?

## 20.2 One-Dimensional Waves

5. A wave pulse travels along a string at a speed of 200 cm/s. What will be the speed if:
   **Note:** Each part below is independent and refers to changes made to the original string.
   a. The string's tension is doubled?

   b. The string's mass is quadrupled (but its length is unchanged)?

   c. The string's length is quadrupled (but its mass is unchanged)?

   d. The string's mass and length are both quadrupled?

6. This is a history graph showing the displacement as a function of time at one point on a string. Did the displacement at this point reach its maximum of 2 mm *before* or *after* the interval of time when the displacement was a constant 1 mm? Explain how you interpreted the graph to answer this question.

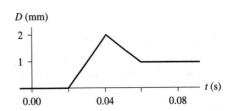

7. Each figure below shows a snapshot graph at time $t = 0$ s of a wave pulse on a string. The pulse on the left is traveling to the right at 100 cm/s; the pulse on the right is traveling to the left at 100 cm/s. Draw snapshot graphs of the wave pulse at the times shown next to the axes.

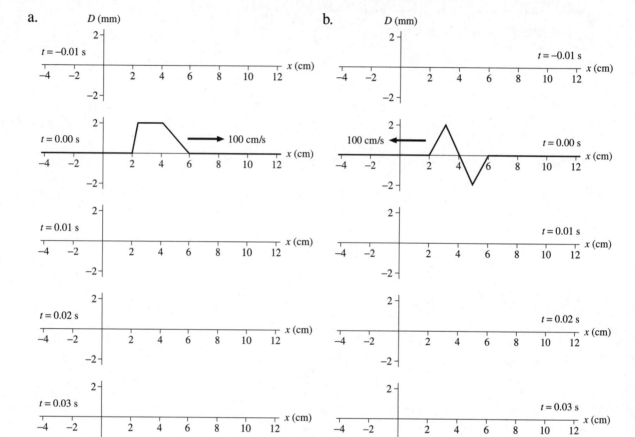

8. This snapshot graph is taken from Exercise 7a. On the axes below, draw the *history* graphs $D(x = 2 \text{ cm}, t)$ and $D(x = 6 \text{ cm}, t)$, showing the displacement at $x = 2$ cm and $x = 6$ cm as functions of time. Refer to your graphs in Exercise 7a to see what is happening at different instants of time.

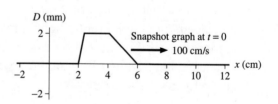

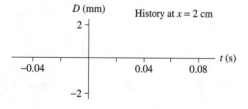

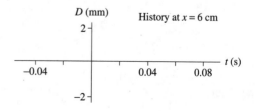

9. This snapshot graph is from Exercise 7b.

   a. Draw the history graph $D(x = 0$ cm, $t)$ for this wave at the point $x = 0$ cm.

   b. Draw the *velocity*-versus-time graph for the piece of the string at $x = 0$ cm. Imagine painting a dot on the string at $x = 0$ cm. What is the velocity of this dot as a function of time as the wave passes by?

   c. As a wave passes through a medium, is the speed of a particle in the medium the same as or different from the speed of the wave through the medium? Explain.

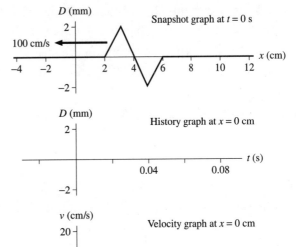

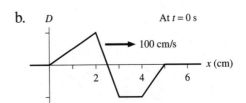

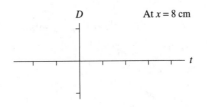

10. Below are four snapshot graphs of wave pulses on a string. For each, draw the history graph at the specified point on the *x*-axis. No time scale is provided on the *t*-axis, so you must determine an appropriate time scale and label the *t*-axis appropriately.

a.
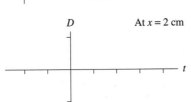

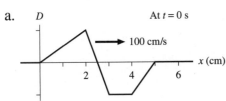

b.
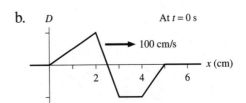

c.
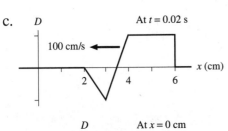

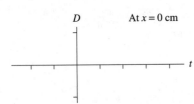

d.
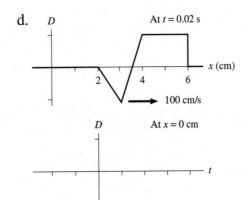

11. A history graph $D(x = 0$ cm$, t)$ is shown for the $x = 0$ cm point on a string. The pulse is moving to the right at 100 cm/s.

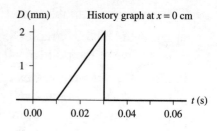

a. Does the $x = 0$ cm point on the string rise quickly and then fall slowly, or rise slowly and then fall quickly? Explain.

b. At what time does the leading edge of the wave pulse arrive at $x = 0$ cm? _____

c. At $t = 0$ s, how far is the leading edge of the wave pulse from $x = 0$ cm? Explain.

d. At $t = 0$ s, is the leading edge to the right or to the left of $x = 0$ cm? _____

e. At what time does the trailing edge of the wave pulse leave $x = 0$ cm? _____

f. At $t = 0$ s, how far is the trailing edge of the pulse from $x = 0$ cm? Explain.

g. By referring to the answers you've just given, draw a snapshot graph $D(x, t = 0$ s$)$ showing the wave pulse on the string at $t = 0$ s.

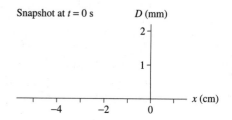

12. These are a history graph *and* a snapshot graph for a wave pulse on a string. They describe the same wave from two perspectives.

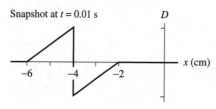 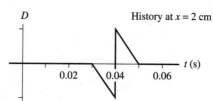

a. In which direction is the wave traveling? Explain.

b. What is the speed of this wave?

13. Below are three history graphs for wave pulses on a string. The speed and direction of each pulse are indicated. For each, draw the snapshot graph at the specified instant of time. No distance scale is provided, so you must determine an appropriate scale and label the *x*-axis appropriately.

a.

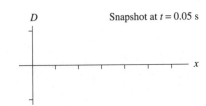

100 cm/s to the right

b.

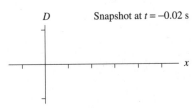

100 cm/s to the left

c.

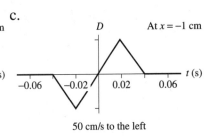

50 cm/s to the left

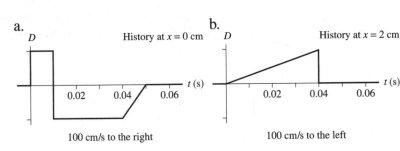

14. A horizontal Slinky is at rest on a table. A wave pulse is sent along the Slinky, causing the top of link 5 to move *horizontally* with the displacement shown in the graph.

a. Is this a transverse or a longitudinal wave? Explain.

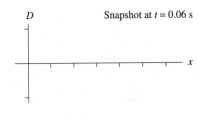

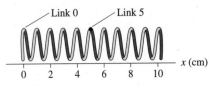

b. What is the position of link 5 at $t = 0.1$ s? _____

What is the position of link 5 at $t = 0.2$ s? _____

What is the position of link 5 at $t = 0.3$ s? _____

**Note:** *Position*, not displacement.

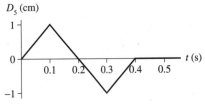

c. Draw a velocity-versus-time graph of link 5. Add an appropriate scale to the vertical axis.

d. Can you determine, from the information given, whether the wave pulse is traveling to the right or to the left? If so, give the direction and explain how you found it. If not, why not?

e. Can you determine, from the information given, the speed of the wave? If so, give the speed and explain how you found it. If not, why not?

15. We can use a series of dots to represent the positions of the links in a Slinky. The top set of dots shows a Slinky in equilibrium with a 1 cm spacing between the links. A wave pulse is sent down the Slinky, traveling to the right at 10 cm/s. The second set of dots shows the Slinky at $t = 0$ s. The links are numbered, and you can measure the displacement $\Delta x$ of each link.

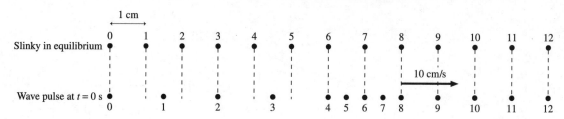

a. Draw a snapshot graph showing the displacement of each link at $t = 0$ s. There are 13 links, so your graph should have 13 dots. Connect your dots with lines to make a continuous graph.

b. Is your graph a "picture" of the wave or a "representation" of the wave? Explain.

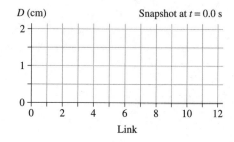

c. Which links are in compression? (list their numbers) _____

Which links are in rarefaction? (list their numbers) _____

d. Draw graphs of displacement versus the link number at $t = 0.1$ s and $t = 0.2$ s.

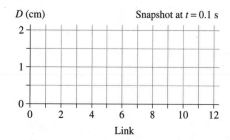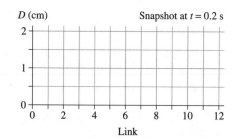

e. Now draw dot pictures of the links at $t = 0.1$ s and $t = 0.2$ s. The equilibrium positions and the $t = 0$ s picture are shown for reference.

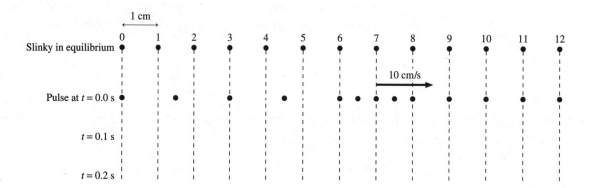

16. The graph shows displacement versus the link number for a wave pulse on a Slinky. Draw a dot picture showing the Slinky at this instant of time. A picture of the Slinky in equilibrium, with 1 cm spacings, is given for reference.

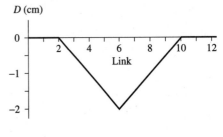

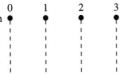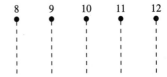

Equilibrium

17. The graph shows displacement versus the link number for a wave pulse on a Slinky. Draw a dot picture showing the Slinky at this instant of time. A picture of the Slinky in equilibrium, with 1 cm spacings, is given for reference.

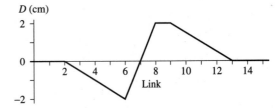

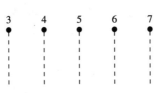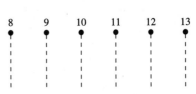

Equilibrium

## 20.3 Sinusoidal Waves

18. The figure shows a sinusoidal traveling wave. Draw a graph of the wave if:

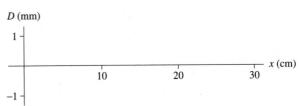

a. Its amplitude is halved and its wavelength is doubled.

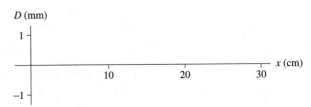

b. Its speed is doubled and its frequency is quadrupled.

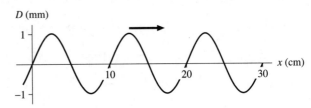

19. The wave shown at time $t = 0$ s is traveling to the right at a speed of 25 cm/s.

    a. Draw snapshot graphs of this wave at times $t = 0.1$ s, $t = 0.2$ s, $t = 0.3$ s, and $t = 0.4$ s.

    b. What is the wavelength of the wave?

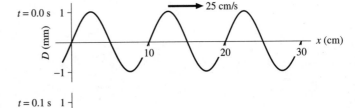

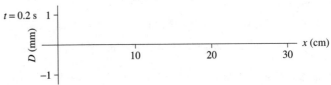

    c. Based on your graphs, what is the period of the wave?

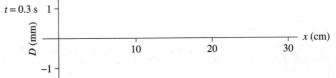

    d. What is the frequency of the wave?

    e. What is the value of the product $\lambda f$?

    f. How does this value of $\lambda f$ compare to the speed of the wave?

20. The phase at one point on a sinusoidal wave is $(3/2)\pi$ rad.

  a. Is the displacement at that point a crest, a trough, zero, or in between? _____

  b. What is the displacement if the phase is $(5/2)\pi$ rad? _____

  c. What is the displacement if the phase is $5\pi$ rad? _____

21. Three waves traveling to the right are shown below. The first two are shown at $t = 0$, the third at $t = T/2$. What are the phase constants $\phi_0$ of these waves?

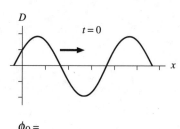 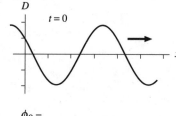 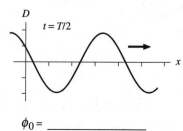

$\phi_0 =$ _____      $\phi_0 =$ _____      $\phi_0 =$ _____

**Note:** Knowing the displacement $D(0,0)$ is a *necessary* piece of information for finding $\phi_0$ but is not by itself enough. The first two waves above have the same value for $D(0,0)$ but they do *not* have the same $\phi_0$. You must also consider the overall shape of the wave.

22. A sinusoidal wave with wavelength 2 m is traveling along the x-axis. At $t = 0$ s the wave's phase at $x = 2$ m is $\pi/2$ rad.

  a. Draw a snapshot graph of the wave at $t = 0$ s.

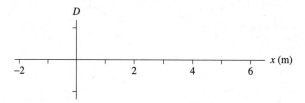

  b. At $t = 0$ s, what is the phase at $x = 0$ m? _____

  c. At $t = 0$ s, what is the phase at $x = 1$ m? _____

  d. At $t = 0$ s, what is the phase at $x = 3$ m? _____

  **Note:** No calculations are needed. Think about what the phase *means* and utilize your graph.

23. Consider the wave shown. Redraw this wave if:

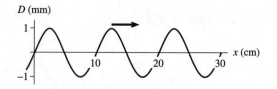

  a. Its wave number is doubled.

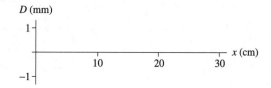

  b. Its wave number is halved.

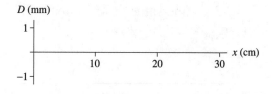

## 20.4 Waves in Two and Three Dimensions

24. A wave-front diagram is shown for a sinusoidal plane wave at time $t = 0$ s. The diagram shows only the $xy$-plane, but the wave extends above and below the plane of the paper.

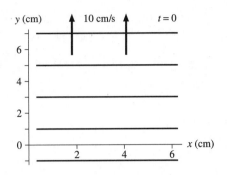

a. What is the wavelength of this wave? _____

b. At $t = 0$ s, for which values of $y$ is the wave a crest?

c. At $t = 0$ s, for which values of $y$ is the wave a trough?

d. Can you tell if this is a transverse or a longitudinal wave? If so, which is it and how did you determine it? If not, why not?

e. How does the displacement at the point $(x, y, z) = (6, 5, 0)$ compare to the displacement at the point $(2, 5, 0)$? Is it more, less, the same, or is there no way to tell? Explain.

f. How does the displacement at the point $(x, y, z) = (2, 5, 5)$ compare to the displacement at the point $(2, 5, 0)$? Is it more, less, the same, or is there no way to tell? Explain.

g. On the left axes below, draw a snapshot graph $D(y, t = 0$ s$)$ along the $y$-axis at $t = 0$ s.

h. On the right axes below, draw a wave-front diagram at time $t = 0.3$ s.

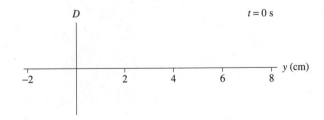

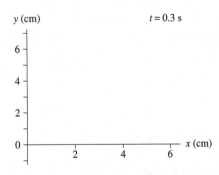

25. These are the wave fronts of a circular wave. What is the phase difference between:

a. Points A and B?    _____

b. Points C and D?    _____

c. Points E and F?    _____

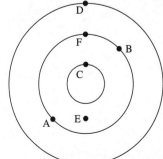

# 20.5 Sound and Light

26. Rank in order, from largest to smallest, the wavelengths $\lambda_1$ to $\lambda_3$ for sound waves having frequencies $f_1 = 100$ Hz, $f_2 = 1000$ Hz, and $f_3 = 10,000$ Hz.

    Order:

    Explanation:

27. A light wave travels from vacuum, through a transparent material, and back to vacuum. What is the index of refraction of this material? Explain.

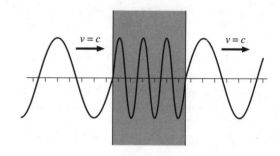

28. A light wave travels from vacuum, through a transparent material whose index of refraction is $n = 2.0$, and back to vacuum. Finish drawing the snapshot graph of the light wave at this instant.

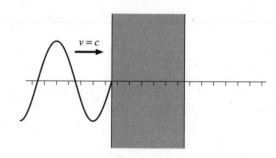

## 20.6  Power and Intensity

29. A laser beam has intensity $I_0$.

    a. What is the intensity, in terms of $I_0$, if a lens focuses the laser beam to $\frac{1}{10}$ its initial diameter?

    b. What is the intensity, in terms of $I_0$, if a lens defocuses the laser beam to 10 times its initial diameter?

30. Sound wave A delivers 2 J of energy in 2 s. Sound wave B delivers 10 J of energy in 5 s. Sound wave C delivers 2 mJ of energy in 1 ms. Rank in order, from largest to smallest, the sound powers $P_A$, $P_B$, and $P_C$ of these three sound waves.

    Order:

    Explanation:

## 20.7  The Doppler Effect

31. You are standing at $x = 0$ m, listening to a sound that is emitted at frequency $f_0$. The graph shows the frequency you hear during a four-second interval. Which of the following describes the sound source?

    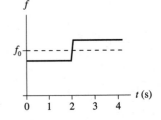

    i.   It moves from left to right and passes you at $t = 2$ s.
    ii.  It moves from right to left and passes you at $t = 2$ s.
    iii. It moves toward you but doesn't reach you. It then reverses direction at $t = 2$ s.
    iv.  It moves away from you until $t = 2$ s. It then reverses direction and moves toward you but doesn't reach you.

    Explain your choice.

32. You are standing at $x = 0$ m, listening to a sound that is emitted at frequency $f_0$. At $t = 0$ s, the sound source is at $x = 20$ m and moving toward you at a steady 10 m/s. Draw a graph showing the frequency you hear from $t = 0$ s to $t = 4$ s. Only the shape of the graph is important, not the numerical values of $f$.

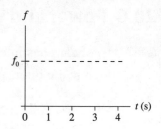

33. You are standing at $x = 0$ m, listening to seven identical sound sources. At $t = 0$ s, all seven are at $x = 343$ m and moving as shown below. The sound from all seven will reach your ear at $t = 1$ s.

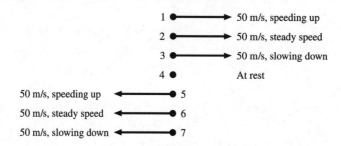

Rank in order, from highest to lowest, the seven frequencies $f_1$ to $f_7$ that you hear at $t = 1$ s.

Order:

Explanation:

# 21 Superposition

## 21.1 The Principle of Superposition

1. Two pulses on a string are approaching each other at 10 m/s. Draw snapshot graphs of the string at the three times indicated.

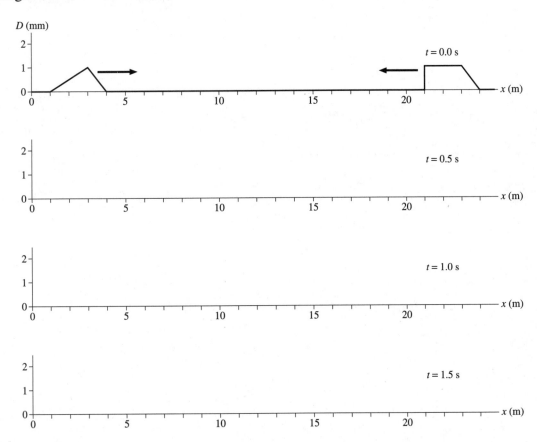

2. Two pulses on a string are approaching each other at 10 m/s. Draw a snapshot graph of the string at $t = 1$ s.

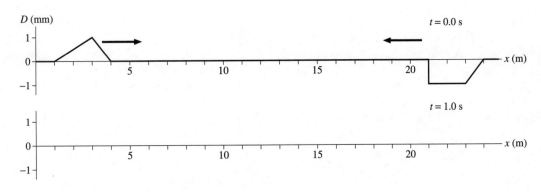

## 21.2 Standing Waves

## 21.3 Transverse Standing Waves

3. Two waves are traveling in opposite directions along a string. Each has a speed of 1000 cm/s, or 1 cm/ms, and an amplitude of 1 cm. The first set of graphs below shows each wave at $t = 0$ ms.

   a. On the axes at the right, draw the superposition of these two waves at $t = 0$ ms.

   b. On the axes at the left, draw each of the two displacements every 1 ms until $t = 8$ ms. The waves extend beyond the graph edges, so new pieces of the wave will move in.

   c. On the axes at the right, draw the superposition of the two waves at the same instant.

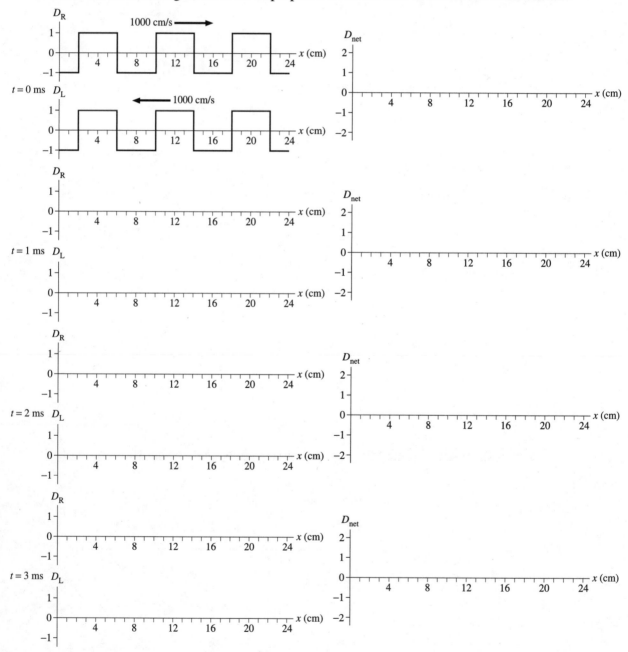

*(Continues next page)*

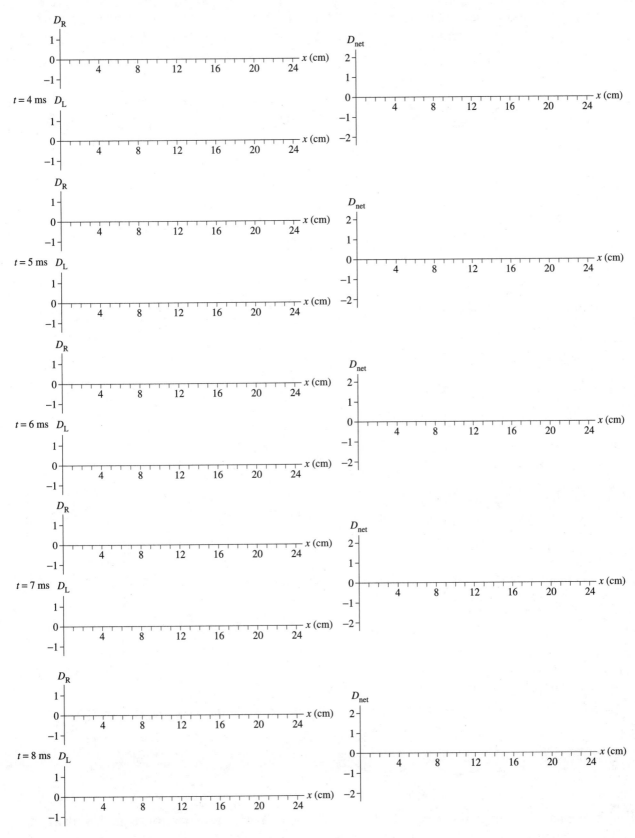

d. What positions on the $x$-axis are nodes? _____

e. What positions on the $x$-axis are antinodes? _____

4. This standing wave has a period of 8 ms. Draw snapshot graphs of the string every 1 ms from *t* = 1 ms to *t* = 8 ms. Think carefully about the proper amplitude at each instant.

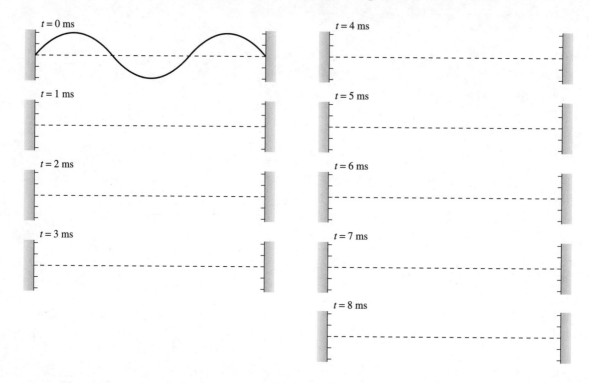

5. The figure shows a standing wave on a string. It has frequency *f*.
   a. Draw the standing wave if the frequency is changed to $\frac{2}{3}f$ and to $\frac{3}{2}f$.

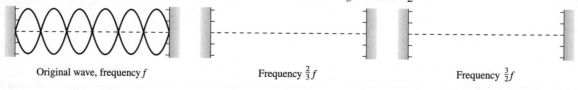

Original wave, frequency *f*        Frequency $\frac{2}{3}f$        Frequency $\frac{3}{2}f$

   b. Is there a standing wave if the frequency is changed to $\frac{1}{4}f$? If so, how many antinodes does it have? If not, why not?

6. The figure shows a standing wave on a string.
   a. Draw the standing wave if the tension is quadrupled while the frequency is held constant.

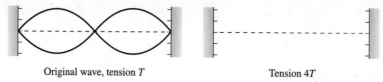

Original wave, tension *T*                Tension 4*T*

   b. Suppose the tension is merely doubled while the frequency remains constant. Will there be a standing wave? If so, how many antinodes will it have? If not, why not?

# 21.4 Standing Sound Waves and Musical Instruments

7. The picture shows a standing sound wave in a 32-mm-long tube of air that is open at both ends.

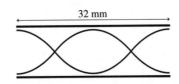

a. Which mode (value of *m*) standing wave is this? _____

b. Are the air molecules vibrating vertically or horizontally? Explain.

c. At what distances from the left end of the tube do the molecules oscillate with maximum amplitude?

8. The purpose of this exercise is to visualize the motion of the air molecules for the standing wave of Exercise 7. On the next page are nine graphs, every one-eighth of a period from $t = 0$ to $t = T$. Each graph represents the displacements at that instant of time of the molecules in a 32-mm-long tube. Positive values are displacements to the right, negative values are displacements to the left.

a. Consider nine air molecules that, in equilibrium, are 4 mm apart and lie along the axis of the tube. The top picture on the right shows these molecules in their equilibrium positions. The dotted lines down the page—spaced 4 mm apart—are reference lines showing the equilibrium positions. Read each graph carefully, then draw nine dots to show the positions of the nine air molecules at each instant of time. The first one, for $t = 0$, has already been done to illustrate the procedure.

**Note:** It's a good approximation to assume that the left dot moves in the pattern 4, 3, 0, −3, −4, −3, 0, 3, 4 mm; the second dot in the pattern 3, 2 , 0, −2, −3, −2, 0, 2, 3 mm; and so on.

b. At what times does the air reach maximum compression, and where does it occur?

Max compression at time _____    Max compression at position _____

_____    _____

_____    _____

c. What is the relationship between the positions of maximum compression and the nodes of the standing wave?

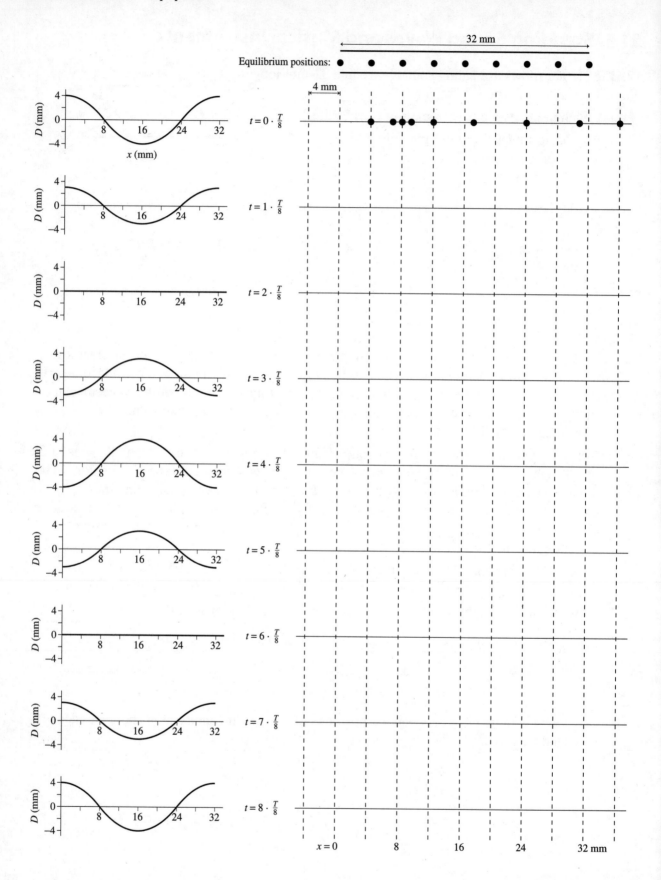

## 21.5  Interference in One Dimension

## 21.6  The Mathematics of Interference

9. The figure shows a snapshot graph at $t = 0$ s of loudspeakers emitting triangular-shaped sound waves. Speaker 2 can be moved forward or backward along the axis. Both speakers vibrate in phase at the same frequency. The second speaker is drawn below the first, so that the figure is clear, but you want to think of the two waves as overlapped as they travel along the $x$-axis.

   a. On the left set of axes, draw the $t = 0$ s snapshot graph of the second wave if speaker 2 is placed at each of the positions shown. The first graph, with $x_{\text{speaker}} = 2$ m, is already drawn.

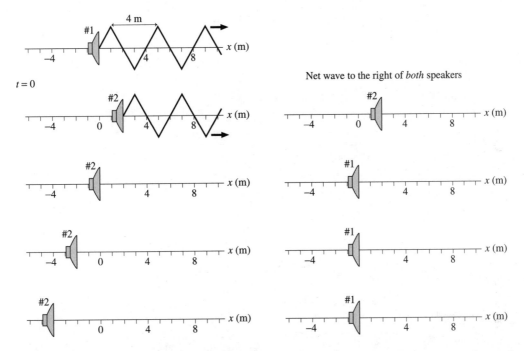

   b. On the right set of axes, draw the superposition $D_{\text{net}} = D_1 + D_2$ of the waves from the two speakers. $D_{\text{net}}$ exists only to the right of *both* speakers. It is the net wave traveling to the right.

   c. What separations between the speakers give constructive interference?  _____

   d. What are the $\Delta x/\lambda$ ratios at the points of constructive interference?  _____

   e. What separations between the speakers give destructive interference?  _____

   f. What are the $\Delta x/\lambda$ ratios at the points of destructive interference?  _____

10. Consider the two loudspeakers of Exercise 9.

   a. Copy the speaker 1 and 2 graphs from Exercise 9 onto the first set of axes below for the situation in which speaker 2 is 4 m behind speaker 1. Then draw their superposition on the axes at the right. This simply repeats your last set of graphs from Exercise 9.

   b. On the axes on the left, draw snapshot graphs of the two waves at times $t = \frac{1}{4}T$, $\frac{2}{4}T$, and $\frac{3}{4}T$, where $T$ is the wave's period.

   c. On the right axes, draw the superposition of the two waves.

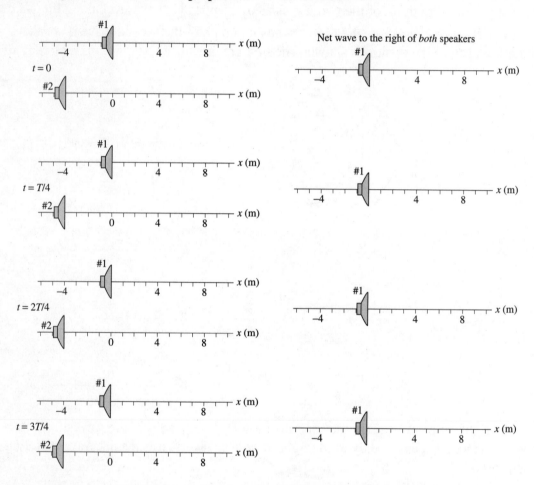

   d. Is the net wave a traveling wave or a standing wave? Use your *observations* to explain.

11. Two loudspeakers are shown at $t = 0$ s. Speaker 2 is 4 m behind speaker 1.

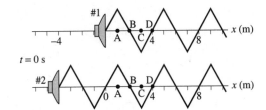

$t = 0$ s

a. Is the interference constructive or destructive?

b. What is the phase constant $\phi_{10}$ for wave 1? _____

   What is the phase constant $\phi_{20}$ for wave 2? _____

c. At points A, B, C, and D on the $x$-axis, what are:
   - The distances $x_1$ and $x_2$ to the two speakers?
   - The path length difference $\Delta x = x_2 - x_1$?
   - The phases $\phi_1$ and $\phi_2$ of the two waves at the point (not the phase constant)?
   - The phase difference $\Delta\phi = \phi_2 - \phi_1$?

   Point A is already filled in to illustrate.

|         | $x_1$ | $x_2$ | $\Delta x$ | $\phi_1$ | $\phi_2$ | $\Delta\phi$ |
|---------|-------|-------|------------|----------|----------|--------------|
| Point A | 1 m   | 5 m   | 4 m        | $0.5\pi$ rad | $2.5\pi$ rad | $2\pi$ rad |
| Point B | _____ | _____ | _____      | _____    | _____    | _____        |
| Point C | _____ | _____ | _____      | _____    | _____    | _____        |
| Point D | _____ | _____ | _____      | _____    | _____    | _____        |

d. Now speaker 2 is placed only 2 m behind speaker 1. Is the interference constructive or destructive?

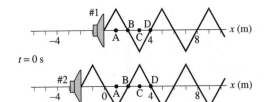

$t = 0$ s

e. Repeat step c for the points A, B, C, and D.

|         | $x_1$ | $x_2$ | $\Delta x$ | $\phi_1$ | $\phi_2$ | $\Delta\phi$ |
|---------|-------|-------|------------|----------|----------|--------------|
| Point A | _____ | _____ | _____      | _____    | _____    | _____        |
| Point B | _____ | _____ | _____      | _____    | _____    | _____        |
| Point C | _____ | _____ | _____      | _____    | _____    | _____        |
| Point D | _____ | _____ | _____      | _____    | _____    | _____        |

f. When the interference is constructive, what is $\Delta x/\lambda$? _____    What is $\Delta\phi$? _____

g. When the interference is destructive, what is $\Delta x/\lambda$? _____    What is $\Delta\phi$? _____

12. Two speakers are placed side-by-side at $x = 0$ m. The waves are shown at $t = 0$ s.

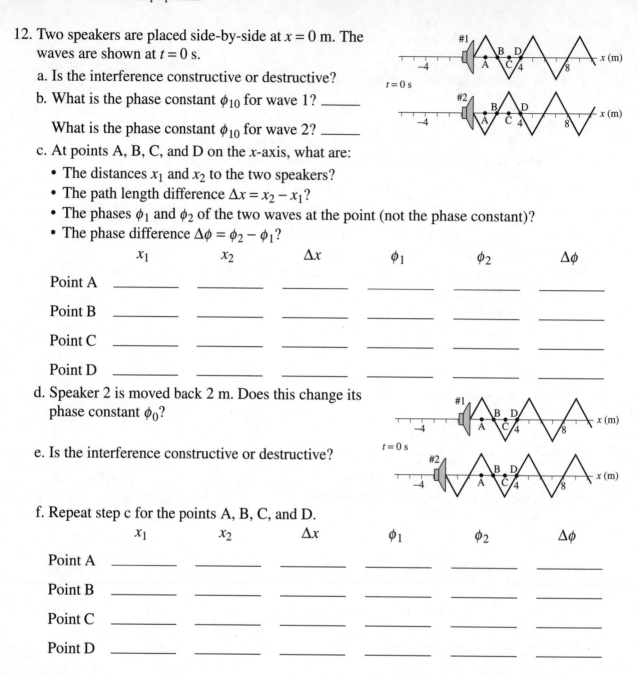

a. Is the interference constructive or destructive?

b. What is the phase constant $\phi_{10}$ for wave 1? _____

What is the phase constant $\phi_{10}$ for wave 2? _____

c. At points A, B, C, and D on the x-axis, what are:
   - The distances $x_1$ and $x_2$ to the two speakers?
   - The path length difference $\Delta x = x_2 - x_1$?
   - The phases $\phi_1$ and $\phi_2$ of the two waves at the point (not the phase constant)?
   - The phase difference $\Delta\phi = \phi_2 - \phi_1$?

|  | $x_1$ | $x_2$ | $\Delta x$ | $\phi_1$ | $\phi_2$ | $\Delta\phi$ |
|---|---|---|---|---|---|---|
| Point A | | | | | | |
| Point B | | | | | | |
| Point C | | | | | | |
| Point D | | | | | | |

d. Speaker 2 is moved back 2 m. Does this change its phase constant $\phi_0$?

e. Is the interference constructive or destructive?

f. Repeat step c for the points A, B, C, and D.

|  | $x_1$ | $x_2$ | $\Delta x$ | $\phi_1$ | $\phi_2$ | $\Delta\phi$ |
|---|---|---|---|---|---|---|
| Point A | | | | | | |
| Point B | | | | | | |
| Point C | | | | | | |
| Point D | | | | | | |

13. Review your answers to the Exercises 9 to 12. Is it the separation path length difference $\Delta x$ or the phase difference $\Delta\phi$ between the waves that determines whether the interference is constructive or destructive? Explain.

# 21.7 Interference in Two and Three Dimensions

14. This is a snapshot graph of two plane waves passing through a region of space. Each has a 2 mm amplitude. At each lettered point, what are the displacements of each wave and the net displacement?

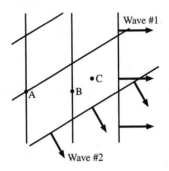

a. Point A: $D_1 = $ _____   $D_2 = $ _____   $D_{net} = $ _____

b. Point B: $D_1 = $ _____   $D_2 = $ _____   $D_{net} = $ _____

c. Point C: $D_1 = $ _____   $D_2 = $ _____   $D_{net} = $ _____

15. Speakers 1 and 2 are 12 m apart. Both emit identical triangular sound waves with $\lambda = 4$ m and $\phi_0 = \pi/2$ rad. Point A is $r_1 = 16$ m from speaker 1.

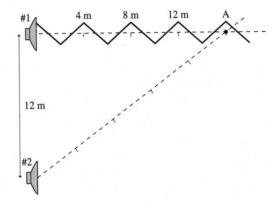

a. What is distance $r_2$ from speaker 2 to A?

b. Draw the wave from speaker 2 along the dashed line to just past point A.

c. At A, is wave 1 a crest, trough, or zero? _____

   At A, is wave 2 a crest, trough, or zero? _____

d. What is the path length difference $\Delta r = r_2 - r_1$? _____   What is the ratio $\Delta r/\lambda$? _____

e. Is the interference at point A constructive, destructive, or in between? _____

16. Speakers 1 and 2 are 18 m apart. Both emit identical triangular sound waves with $\lambda = 4$ m and $\phi_0 = \pi/2$ rad. Point B is $r_1 = 24$ m from speaker 1.

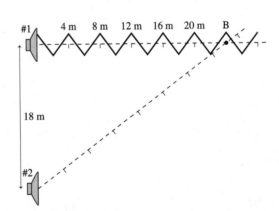

a. What is distance $r_2$ from speaker 2 to B?

b. Draw the wave from speaker 2 along the dashed line to just past point A.

c. At B, is wave 1 a crest, trough, or zero? _____

   At B, is wave 2 a crest, trough, or zero? _____

d. What is the path length difference $\Delta r = r_2 - r_1$? _____   What is the ratio $\Delta r/\lambda$? _____

e. Is the interference at point B constructive, destructive, or in between? _____

17. Two speakers 12 m apart emit identical triangular sound waves with $\lambda = 4$ m and $\phi_0 = 0$ rad. The distances $r_1$ to points A, B, C, D, and E are shown in the table below.

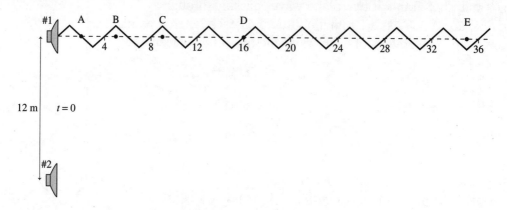

a. For each point, fill in the table and determine whether the interference is constructive (C) or destructive (D).

| Point | $r_1$ | $r_2$ | $\Delta r$ | $\Delta r/\lambda$ | C or D |
|---|---|---|---|---|---|
| A | 2.2 m | | | | |
| B | 5.0 m | | | | |
| C | 9.0 m | | | | |
| D | 16 m | | | | |
| E | 35 m | | | | |

b. Are there any points to the right of E, on the line straight out from speaker 1, for which the interference is either exactly constructive or exactly destructive? If so, where? If not, why not?

c. Suppose you start at speaker 1 and walk straight away from the speaker for 50 m. Describe what you will hear as you walk.

18. The figure shows the wave-front pattern emitted by two loudspeakers.

   a. Draw a dot • at points where there is constructive interference. These will be points where two crests overlap *or* two troughs overlap.

   b. Draw an open circle ∘ at points where there is destructive interference. These will be points where a crest overlaps a trough.

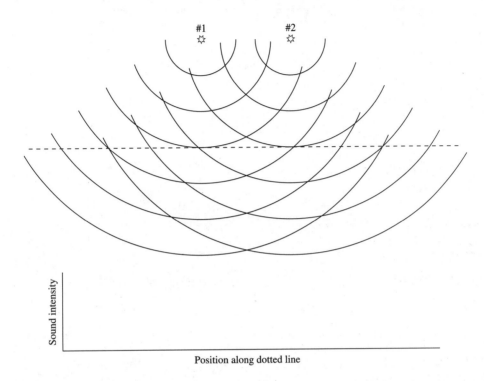

c. Use a **black** line to draw each "ray" of constructive interference. Use a **red** line to draw each "ray" of destructive interference.

d. Draw a graph on the axes above of the sound intensity you would hear if you walked along the horizontal dotted line. Use the same horizontal scale as the figure so that your graph lines up with the figure above it.

e. Suppose the phase constant of speaker 2 is increased by $\pi$ rad. Describe what will happen to the interference pattern.

## 21.8 Beats

19. The two waves arrive simultaneously at a point in space from two different sources.

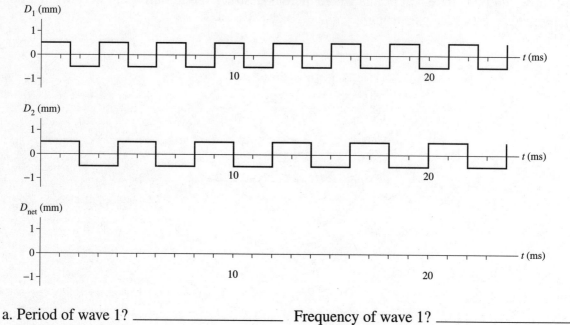

a. Period of wave 1? _____ Frequency of wave 1? _____

b. Period of wave 2? _____ Frequency of wave 2? _____

c. Draw the graph of the net wave at this point on the third set of axes. Be accurate, use a ruler!

d. Period of the net wave? _____ Frequency of the net wave? _____

e. Is the frequency of the superposition what you would expect as a beat frequency? Explain.

# 22 Wave Optics

## 22.1 Light and Optics

## 22.2 The Interference of Light

1. The figure shows the light intensity recorded by a piece of film in an interference experiment. Notice that the light intensity comes "full on" at the edges of each maximum, so this is *not* the intensity that would be recorded in Young's double-slit experiment.

   a. Draw a graph of light intensity versus position on the film. Your graph should have the same horizontal scale as the "photograph" above it.

   b. Is it possible to tell, from the information given, what the wavelength of the light is? If so, what is it? If not, why not?

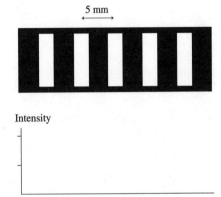

2. The graph shows the light intensity on the viewing screen during a double-slit interference experiment.

   a. Draw the "photograph" that would be recorded if a piece of film were placed at the position of the screen. Your "photograph" should have the same horizontal scale as the graph above it. Be as accurate as you can. Let the white of the paper be the brightest intensity and a very heavy pencil shading be the darkest.

   b. Three positions on the screen are marked as A, B, and C. Draw history graphs showing the displacement of the light wave at each of these three positions as a function of time. Show three cycles, and use the same vertical scale on all three.

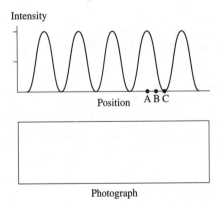

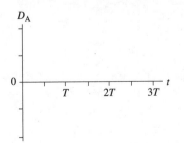

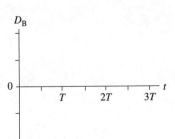

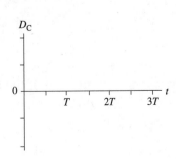

3. In a double-slit experiment, we usually see the light intensity only on a viewing screen. However, we can use smoke or dust to make the light visible as it propagates between the slits and the screen. Consider a double-slit experiment in a smoke-filled room. What kind of light and dark pattern would you see if you looked down on the experiment from above? Draw the pattern on the figure below. Shade the areas that are dark and leave the white of the paper for the areas that are bright.

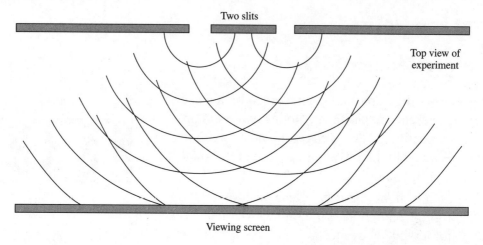

**Exercises 4–7:** Refer to this figure of the viewing screen in a double-slit experiment.

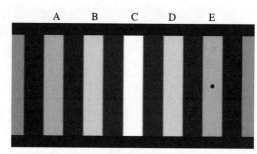

4. What will happen to the fringe spacing if the wavelength of the light is decreased?

5. What will happen to the fringe spacing if the spacing between the slits is decreased?

6. What will happen to the fringe spacing if the distance to the screen is decreased?

7. Suppose the wavelength of the light is 500 nm. How much further is it from the dot in the center of fringe E to the more distant slit than it is from the dot to the nearer slit?

## 22.3  The Diffraction Grating

8. The figure shows four slits in a diffraction grating. A set of Huygens wavelets is spreading out from each slit. Four wave paths, numbered 1 to 4, are shown leaving the slits at angle $\theta_1$. The dotted lines are drawn perpendicular to the paths of the waves.

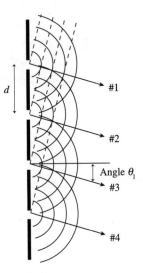

   a. Use a colored pencil or heavy shading to show *on the figure* the extra distance traveled by wave 1 that is not traveled by wave 2.

   b. How many extra wavelengths does wave 1 travel compared to wave 2? Explain how you can tell from the figure.

   c. How many extra wavelengths does wave 2 travel compared to wave 3?

   d. As these four waves combine at some large distance from the grating, will they interfere constructively, destructively, or in between? Explain.

9. Suppose the wavelength of the light in Exercise 8 is doubled. (Imagine erasing every other wave front in the picture.) Would the interference at angle $\theta_1$ then be constructive, destructive, or in between? Explain. Your explanation should be based on the figure, not on some equation.

10. Suppose the slit spacing $d$ in Exercise 8 is doubled while the wavelength is unchanged. Would the interference at angle $\theta_1$ then be constructive, destructive, or in between? Again, base your explanation on the figure.

11. These are the same slits as in Exercise 8. Waves with the same wavelength are spreading out on the right side.

   a. Draw four paths, starting at the slits, at an angle $\theta_2$ such that the wave along each path travels *two* wavelengths farther than the next wave. Also draw dashed lines at right angles to the travel direction. Your picture should look much like the figure of Exercise 8, but with the waves traveling at a different angle. Use a ruler!

   b. Do the same for four paths at angle $\theta_{1/2}$ such that each wave travels *one-half* wavelength farther than the next wave.

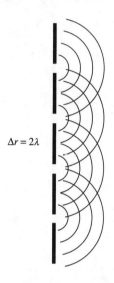

$\Delta r = 2\lambda$

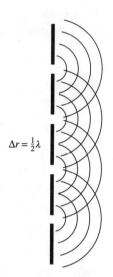

$\Delta r = \frac{1}{2}\lambda$

12. This is the interference pattern on a viewing screen behind two slits. How would the pattern change if the two slits were replaced by 20 slits having the *same spacing d* between adjacent slits?

   a. Would the number of fringes on the screen increase, decrease, or stay the same?

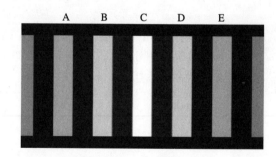

   b. Would the fringe spacing increase, decrease, or stay the same?

   c. Would the width of each fringe increase, decrease, or stay the same?

   d. Would the brightness of each fringe increase, decrease, or stay the same?

## 22.4  Single-Slit Diffraction

13. Plane waves of light are incident on two narrow, closely-spaced slits. The graph shows the light intensity seen on a screen behind the slits.

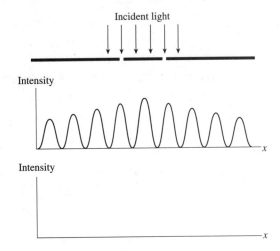

Incident light

a. Draw a graph on the axes below to show the light intensity on the screen if the right slit is blocked, allowing light to go only through the left slit.

b. Explain why the graph will look this way.

14. This is the light intensity on a viewing screen behind a slit of width $a$. The light's wavelength is $\lambda$. Is $\lambda < a$, $\lambda = a$, $\lambda > a$, or is it not possible to tell? Explain.

15. This is the light intensity on a viewing screen behind a rectangular opening in a screen. Is the shape of the opening

☐ or ▯ or □ ?

Explain.

16. The graph shows the light intensity on a screen behind a 0.2-mm-wide slit illuminated by light with a 500 nm wavelength.

   a. Draw a *picture* in the box of how a photograph taken at this location would look. Use the same horizontal scale, so that your picture aligns with the graph above. Let the white of the paper represent the brightest intensity and the darkest you can draw with a pencil or pen be the least intensity.

   b. Using the same horizontal scale as in part a, draw graphs showing the light intensity if

      i.   $\lambda = 250$ nm, $a = 0.2$ mm.

      ii.  $\lambda = 1000$ nm, $a = 0.2$ mm.

      iii. $\lambda = 500$ nm, $a = 0.1$ mm.

      iv.  $\lambda = 500$ nm, $a = 0.4$ mm.

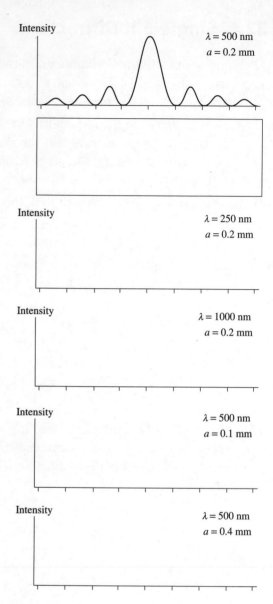

## 22.5 Circular-Aperture Diffraction

17. This is the light intensity on a viewing screen behind a circular aperture. What happens to the width of the central maximum if

   a. The wavelength is increased?

   b. The diameter of the aperture is increased?

   c. How will the screen appear if the aperture diameter is less than the light wavelength?

# 22.6 Interferometers

18. The figure shows a tube through which sound waves with $\lambda = 4$ cm travel from left to right. Each wave divides at the first junction and recombines at the second. The dots and triangles show the positions of the wave crests at $t = 0$ s—rather like a very simple wave front diagram.

    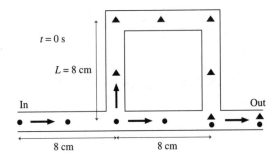

    a. Do the recombined waves interfere constructively or destructively? Explain.

    b. How much *extra* distance does the upper wave travel?  _____

    How many wavelengths is this extra distance?  _____

    c. Below are tubes with $L = 9$ cm and $L = 10$ cm. Use dots to show the wave crest positions at $t = 0$ s for the wave taking the lower path. Use triangles to show the wave crests at $t = 0$ s for the wave taking the upper path. The wavelength is $\lambda = 4$ cm. Assume that the first crest is at the left edge of the tube, as in the figure above.

    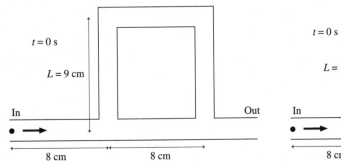 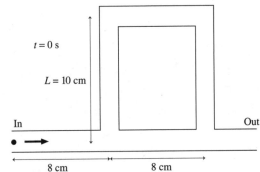

    d. How many *extra* wavelengths does the upper wave travel in the $L = 9$ cm tube?  _____

    What kind of interference does the $L = 9$ cm tube produce?  _____

    e. How many *extra* wavelengths does the upper wave travel in the $L = 10$ cm tube?  _____

    What kind of interference does the $L = 10$ cm tube produce?  _____

# 23 Ray Optics

**Note:** Please use a ruler or straight edge for drawing light rays.

## 23.1 The Ray Model of Light

1. If you turn on your car headlights during the day, the road ahead of you doesn't appear to get brighter. Why not?

2. a. Draw four or five rays from the object that allow A to see the object.
   b. Draw four or five rays from the object that allow B to see the object.

A

C

B

   c. Describe the situations seen by A and B if a piece of cardboard is lowered at point C.

3. a. Draw four or five rays from object 1 that allow A to see object 1.
   b. Draw four or five rays from object 2 that allow B to see object 2.
   c. What happens to the light where the rays cross in the center of the picture?

1 ☆

B

2 ☆

A

4. A point source of light illuminates a slit in a opaque barrier.

   a. On the screen, sketch the pattern of light that you expect to see. Let the white of the paper represent light areas; shade dark areas. Mark any relevant dimensions.

   b. What will happen to the pattern of light on the screen if the slit width is reduced to 0.5 cm?

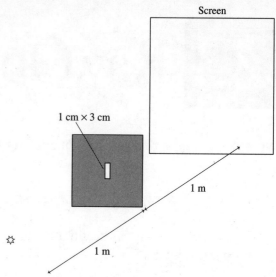

5. In each situation below, light passes through a 1-cm-diameter hole and is viewed on a screen. For each, sketch the pattern of light that you expect to see on the screen. Let the white of the paper represent light arcas; shade dark areas.

a.    b.    c.

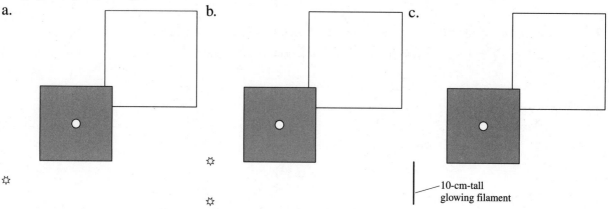

6. Light from a bulb passes through a pinhole. On the screen, sketch the pattern of light that you expect to see.

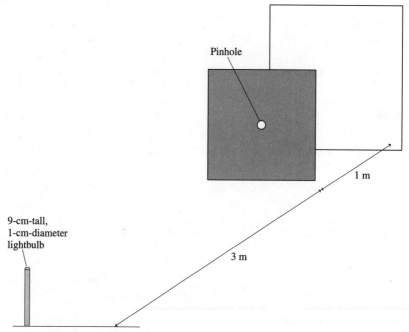

# 23.2 Reflection

7. a. Draw five rays from the object that pass through points A to E after reflecting from the mirror. Make use of the grid to do this accurately.

   b. Extend the reflected rays behind the mirror.

   c. Show and label the image point.

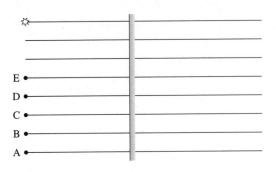

8. a. Draw *one* ray from the object that enters the eye after reflecting from the mirror.

   b. Is one ray sufficient to tell your eye/brain where the image is located?

   c. Use a different color pen or pencil to draw two more rays that enter the eye after reflecting. Then use the three rays to locate (and label) the image point.

   d. Do any of the rays that enter the eye actually pass through the image point?

9. You are looking at the image of a pencil in a mirror.

   a. What happens to the image if the top half of the mirror, down to the midpoint, is covered with a piece of cardboard? Explain.

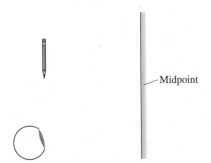

   b. What happens to the image if the bottom half of the mirror is covered with a piece of cardboard? Explain.

10. The two mirrors are perpendicular to each other.

   a. Draw a ray directly from the object to point A. Then draw two rays that strike the mirror *very close* to A, one on either side. Use the reflections of these three rays to locate an image point.

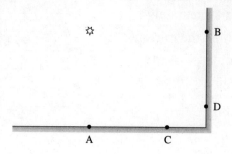

   b. Do the same for points B, C, and D.

   c. How many images are there, and where are they located?

## 23.3 Refraction

11. Draw seven rays from the object that refract after passing through the seven dots on the boundary.

a.

$n_1 < n_2$

b.

$n_1 = n_2$

c.

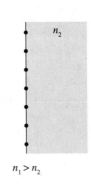

$n_1 > n_2$

12. Complete the trajectories of these three rays through material 2 and back into material 1. Assume $n_2 < n_1$.

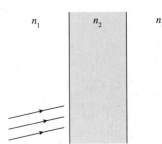

13. The figure shows six conceivable trajectories of light rays leaving an object. Which, if any, of these trajectories are impossible? For each that is possible, what are the requirements of the index of refraction $n_2$?

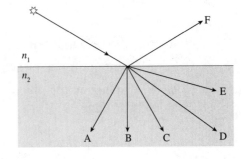

Impossible _____

Requires $n_2 > n_1$ _____

Requires $n_2 = n_1$ _____

Requires $n_2 < n_1$ _____

Possible for any $n_2$ _____

14. Complete the ray trajectories through the two prisms shown below.

a.

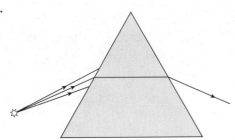

b.

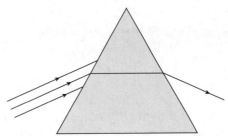

15. Draw the trajectories of seven rays that leave the object heading toward the seven dots on the boundary. Assume $n_2 < n_1$ and $\theta_c = 45°$.

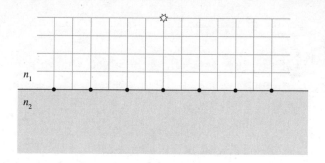

## 23.4 Image Formation by Refraction

16. a. Draw rays that refract after passing through points B, C, and D. Assume $n_2 > n_1$.

   b. Use dotted lines to extend these rays backward into medium 1. Locate and label the image point.

   c. Now draw the rays that refract at A and E.

   d. Use a different color pen or pencil to draw three rays from the object that enter the eye.

   e. Does the distance to the object *appear* to be larger than, smaller than, or the same as the true distance? Explain.

17. A thermometer is partially submerged in an aquarium. The underwater part of the thermometer is not shown.

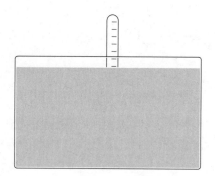

   a. As you look at the thermometer, does the underwater part appear to be closer than, farther than, or the same distance as the top of the thermometer?

   b. Complete the drawing by drawing the bottom of the thermometer as you think it would look.

# 23.5  Color and Dispersion

18. A beam of white light from a flashlight passes through a red piece of plastic.

    a. What is the color of the light that emerges from the plastic?

    b. Is the emerging light as intense as, more intense than, or less intense than the white light?

    c. The light then passes through a blue piece of plastic. Describe the color and intensity of the light that emerges.

19. Suppose you looked at the sky on a clear day through pieces of red and blue plastic oriented as shown. Describe the color and brightness of the light coming through sections 1, 2, and 3.

    Section 1:

    Section 2:

    Section 3:

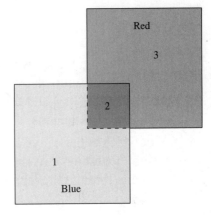

20. Sketch a plausible absorption spectrum for a patch of bright red paint.

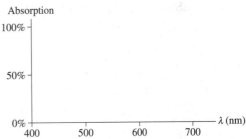

21. The center of the galaxy is filled with hydrogen gas. The density is very low, but the distances are vast. An astronomer wants to take a picture of the center of the galaxy. Will the view be better using ultraviolet light, visible light, or infrared light? (High quality telescopes and cameras are available in all three spectral regions.) Explain the reason for your choice.

## 23.6 Thin Lenses: Ray Tracing

22. a. Continue these rays through the lens and out the other side.

    b. Is the point where the rays converge the same as the focal point of the lens? Or different? Explain.

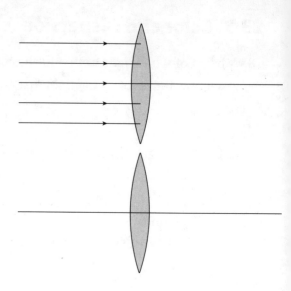

    c. Place a point source of light at the place where the rays converged in part b. Draw several rays heading left, toward the lens. Continue the rays through the lens and out the other side.

    d. Do these rays converge? If so, where?

23. The top two figures show test data for a lens. The third figure shows a point source near this lens and four rays heading toward the lens.

    a. For which of these rays do you know, from the test data, its direction after passing through the lens?

    b. Draw the rays you identified in part a as they pass through the lens and out the other side.

    c. Use a different color pen or pencil to draw the trajectories of the other rays.

    d. Label the image point. What kind of image is this?

    e. The fourth figure shows a second point source. Use ray tracing to locate its image point.

    f. The fifth figure shows an extended object. Have you learned enough to locate its image? If so, draw it.

    g. The last figure shows another extended object. Draw its image.

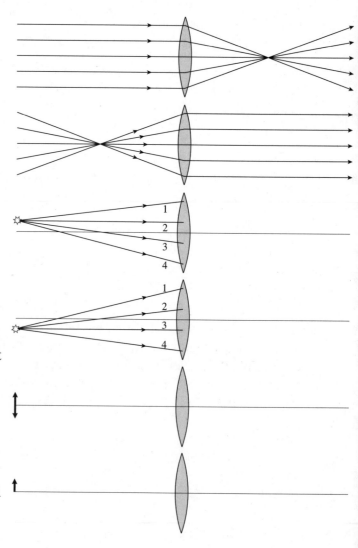

24. An object is near a lens whose focal points are marked with dots.

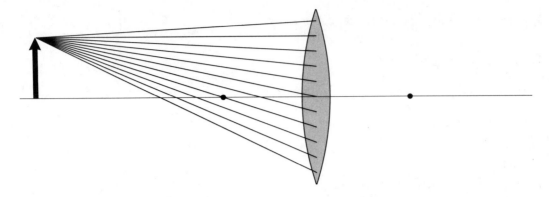

a. Identify the three special rays and continue them through the lens.

b. Use a different color pen or pencil to draw the trajectories of the other rays.

25. a. Consider *one* point on an object near a lens. What is the minimum number of rays needed to locate its image point?

b. For each point on the object, how many rays from this point actually strike the lens and refract to the image point?

26. An object and lens are positioned to form a well-focused, inverted image on a viewing screen. Then a piece of cardboard is lowered just in front of the lens to cover the *top half* of the lens. Describe what happens to the image on the screen. What will you see when the cardboard is in place?

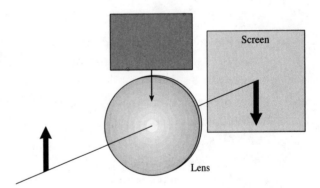

27. An object is near a lens whose focal points are shown.

    a. Use ray tracing to locate the image of this object.

    b. Is the image upright or inverted?

    c. Is the image height larger or smaller than the object height?

    d. Is this a real or a virtual image? Explain how you can tell.

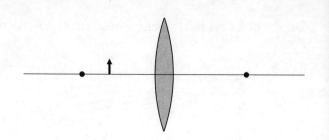

28. The top two figures show test data for a lens. The third figure shows a point source near this lens and four rays heading toward the lens.

    a. For which of these rays do you know, from the test data, its direction after passing through the lens?

    b. Draw the rays you identified in part a as they pass through the lens and out the other side.

    c. Use a different color pen or pencil to draw the trajectories of the other rays.

    d. Find and label the image point. What kind of image is this?

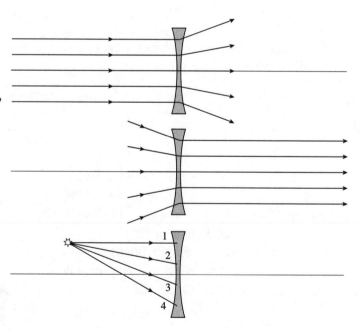

# 23.7 Thin Lenses: Refraction Theory

29. Materials 1 and 2 are separated by a spherical surface. For each part:
    i.   Draw the normal to the surface at the seven dots on the boundary.
    ii.  Draw the trajectories of seven rays from the object that pass through the seven dots.
    iii. Trace the refracted rays either forward to a point where they converge or backward to a point from which they appear to diverge.

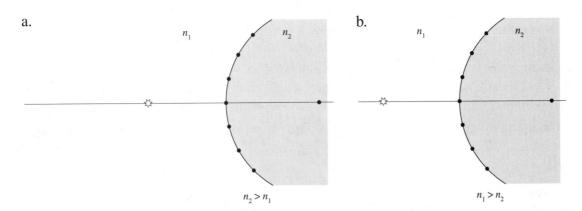

30. Two glass lenses are shown below. They are surrounded by air.
    i.   Draw the normals to the left surface at the seven dots on the boundary.
    ii.  Draw the trajectories of seven rays from the object through the seven dots. Stop when the rays reach the right surface of the lens.
    iii. Draw the normals to the right surface at the seven points reached by the rays.
    iv.  Complete the ray trajectories back into the air on the right side of the lens.

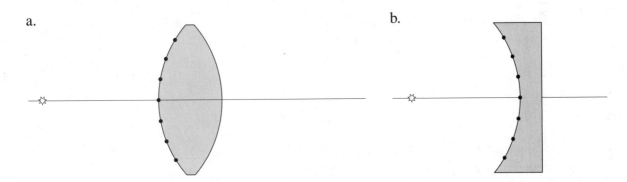

c. Is the part-a lens a converging or a diverging lens? _____

   Is the part-b lens a converging or a diverging lens? _____

## 23.8 The Resolution of Optical Instruments

31. A diffraction-limited lens can focus light to a 10-$\mu$m-diameter spot on a screen. Do the following actions make the spot diameter larger, smaller, or leave it unchanged?

    a. Decreasing the wavelength of the light: _____

    b. Decreasing the lens diameter: _____

    c. Decreasing the lens focal length: _____

    d. Decreasing the lens-to-screen distance: _____

32. An astronomer is trying to observe two distant stars. The stars are marginally resolved when she looks at them through a filter that passes green light near 550 nm. Which of the following actions would improve the resolution? Assume that the resolution is not limited by the atmosphere.

    a. Changing the filter to a different wavelength? If so, should she use a shorter or a longer wavelength?

    b. Using a telescope with an objective lens of the same diameter but a different focal length? If so, should she select a shorter or a longer focal length?

    c. Using a telescope with an objective lens of the same focal length but a different diameter? If so, should she select a larger or a smaller diameter?

    d. Using an eyepiece with a different magnification? If so, should she select an eyepiece with more or less magnification?

# 24 Modern Optics and Matter Waves

## 24.1 Spectroscopy: Unlocking the Structure of Atoms

## 24.2 X-Ray Diffraction

## 24.3 Photons

1. The figure shows the spectrum of a gas discharge tube.

400 nm           500 nm           600 nm           700 nm

What color would the discharge appear to your eye? Explain.

2. The first-order x-ray diffraction of monochromatic x rays from a crystal occurs at angle $\theta_1$. The crystal is then compressed, causing a slight reduction in its volume. Does $\theta_1$ increase, decrease, or stay the same? Explain.

3. Three laser beams have wavelengths $\lambda_1 = 400$ nm, $\lambda_2 = 600$ nm, and $\lambda_3 = 800$ nm. The power of each laser beam is 1 W.

   a. Rank in order, from largest to smallest, the photon energies $E_1$, $E_2$ and $E_3$ in these three laser beams.

   Order:

   Explanation:

b. Rank in order, from largest to smallest, the number of photons per second $N_1$, $N_2$, and $N_3$ delivered by the three laser beams.

Order:

Explanation:

4. The top figure is the *negative* of the photograph of a single-slit diffraction pattern. That is, the darkest areas in the figure were the brightest areas on the screen. This photo was made with an extremely large number of photons.

Suppose the slit is illuminated by an extremely weak light source, so weak that only 1 photon passes through the slit every second. Data are collected for 60 seconds. Draw 60 dots on the empty screen to show how you think the screen will look after 60 photons have been detected.

5. A light source at point A emits light with a wavelength of 1.0 $\mu$m. One photon of light is detected at point B, 5.0 $\mu$m away from A. On the figure, draw the trajectory that the photon follows from A to B.

A
☆

B
•

5 $\mu$m

## 24.4 Matter Waves

6. The figure is a simulation of the electrons detected behind a very narrow double slit. Each bright dot represents one electron. How will this pattern change if the following experimental conditions are changed? Possible changes you should consider include the number of dots and the spacing, width, and positions of the fringes.

   a. The electron-beam intensity is increased.




   b. The electron speed is reduced.




   c. The electrons are replaced by positrons with the same speed. Positrons are antimatter particles that are identical to electrons except that they have a positive charge.




   d. One slit is closed.




7. Very slow neutrons pass through a single, very narrow slit. Use 50 or 60 dots to show how the neutron intensity will appear on a neutron-detector screen behind the slit.

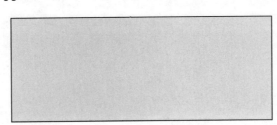

8. To have the best resolution, should an electron microscope use very fast electrons or very slow electrons? Explain.

## 24.5 Energy Is Quantized

9. a. For the allowed energies of a particle in a box to be large, should the box be very big or very small? Explain.

   b. Which is likely to have larger values for the allowed energies: an atom in a molecule, an electron in an atom, or a proton in a nucleus? Explain.

10. The smallest allowed energy of a hydrogen atom (atomic mass number 1) in a box of length $L_0$ is $1.0 \times 10^{-20}$ J. What is the smallest allowed energy of a helium atom (atomic mass number 4) in a box of length $\frac{1}{2}L_0$?

# 25 Electric Charges and Forces

## 25.1 Developing a Charge Model

**Exercises 1–11:** Your answers in Section 25.1 should make *no* mention of electrons or protons.

1. What is alike about charges when we say "two like charges?" Do they look, smell, or taste the same? Base your answer on experimental procedures and observations.

2. Can an insulator be charged? If so, how would you charge an insulator? If not, why not?

3. Can a conductor be charged? If so, how would you charge a conductor? If not, why not?

4. Two lightweight balls hang straight down when both are neutral. They are close enough together to interact, but not close enough to touch. Draw pictures showing how the balls hang if:
   a. Both are touched with a plastic rod that was rubbed with wool.

A        B

b. The two charged balls of part a are moved farther apart.

c. Ball A is touched by a plastic rod that was rubbed with wool and ball B is touched by a glass rod that was rubbed with silk.

d. Both are charged by a plastic rod, but ball A is charged more than ball B.

e. Ball A is charged by a plastic rod. Ball B is neutral.

f. Ball A is charged by a glass rod. Ball B is neutral.

5. Four lightweight balls A, B, C, and D are suspended by threads. Ball A has been touched by a plastic rod that was rubbed with wool. When the balls are brought close together, without touching, the following observations are made:
   - Balls B, C, and D are attracted to ball A.
   - Balls B and D have no effect on each other.
   - Ball B is attracted to ball C.

   What are the charge states (glass, plastic, or neutral) of balls A, B, C, and D? Explain.

6. Charged plastic and glass rods hang by threads.

   a. An object repels the plastic rod. Can you predict what it will do to the glass rod? If so, what? If not, why not? Explain.

   b. A different object attracts the plastic rod. Can you predict what it will do to the glass rod? If so, what? If not, why not? Explain.

7. After combing your hair briskly, the comb will pick up small pieces of paper.

   a. Is the comb charged? Explain.

   b. How can you be sure that it isn't the paper that is charged? Propose an experiment to test this.

   c. Is your hair charged after being combed? What evidence do you have for your answer?

   d. What kind of charge is the comb likely to have? Why?

   e. How could you test your answer to part d?

8. When you take clothes out of the drier right after it stops, the clothes often stick to your hands and arms. Is your body charged? If so, how did it acquire a charge? If not, why does this happen?

9. A lightweight metal ball hangs by a thread. When a charged rod is held near, the ball moves toward the rod, touches the rod, then quickly "flies away" from the rod. Explain this behavior.

10. You've been given a piece of material. Propose an experiment or a series of experiments to determine if the material is a conductor or an insulator. State clearly what the outcome of each experiment will be if the material is a conductor and if it is an insulator.

11. Suppose there exists a third type of charge in addition to the two types we've called glass and plastic. Call this third type X charge. What experiment or series of experiments would you use to test whether an object has X charge? State clearly how each possible outcome of the experiments is to be interpreted.

## 25.2 Charge

## 25.3 Insulators and Conductors

12. A negatively charged electroscope has separated leaves.

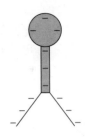

    a. Suppose you bring a negatively charged rod close to the top of the electroscope, but not touching. How will the leaves respond? Use both charge diagrams and words to explain.

    b. How will the leaves respond if you bring a positively charged rod close to the top of the electroscope, but not touching? Use both charge diagrams and words to explain.

13. a. A negatively charged plastic rod touches a neutral piece of metal. What is the final charge state (positive, negative, or neutral) of the metal? Use both charge diagrams and words to explain how this charge state is achieved.

    b. A positively charged glass rod touches a neutral piece of metal. What is the final charge state of the metal? Use both charge diagrams and words to explain how this charge state is achieved.

14. A lightweight, positively charged ball and a neutral rod hang by threads. They are close but not touching. A positively charged glass rod touches the hanging rod on the end opposite the ball, then the rod is removed.

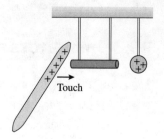

a. Draw a picture of the final positions of the hanging rod and the ball if the rod is made of glass.

b. Draw a picture of the final positions of the hanging rod and the ball if the rod is metal.

15. Two oppositely charged metal spheres have equal quantities of charge. They are brought into contact with a neutral metal rod.

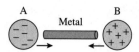

a. What is the final charge state of each sphere and of the rod?

b. Give a microscopic explanation, in terms of fundamental charges, of how these final states are reached. Use both charge diagrams and words.

16. Metal sphere A has 4 units of negative charge and metal sphere B has 2 units of positive charge. The two spheres are brought into contact. What is the final charge state of each sphere? Explain.

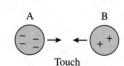

17. a. Metal sphere A is initially neutral. A positively charged rod is brought near, but not touching. Is A now positive, negative, or neutral? Explain.

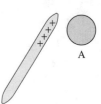

b. Metal spheres A and B are initially neutral and are touching. A positively charged rod is brought near A, but not touching. Is A now positive, negative, or neutral? Explain.

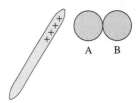

c. Metal sphere A is initially neutral. It is connected by a metal wire to the ground. A positively charged rod is brought near, but not touching. Is A now positive, negative, or neutral? Explain.

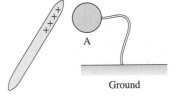

18. A lightweight, positively charged ball and a neutral metal rod hang by threads. They are close but not touching. A positively charged rod is held close to, but not touching, the hanging rod on the end opposite the ball.

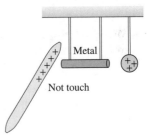

a. Draw a picture of the final positions of the hanging rod and the ball. Explain your reasoning.

b. Suppose the positively charged rod is replaced with a negatively charged rod. Draw a picture of the final positions of the hanging rod and the ball. Explain your reasoning.

19. A positively charged rod is held near, but not touching, a neutral metal sphere.

    a. Add plusses and minuses to the figure to show the charge distribution on the sphere.

    b. Does the sphere experience a net force? If so, in which direction? Explain.

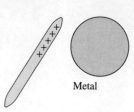

Metal

20. If you bring your finger near a lightweight, negatively charged hanging ball, the ball swings over toward your finger. Use charge diagrams and words to explain this observation.

Finger

21. The figure shows an atom with four protons in the nucleus and four electrons in the electron cloud.

    a. Draw a picture showing how this atom will look if a positive charge is held just *above* the atom.

    b. Is there a net force on the atom? If so, in which direction? Explain.

# 25.4 Coulomb's Law

22. For each pair of charges, draw a force vector *on each charge* to show the electric force acting on that charge. The length of each vector should be proportional to the magnitude of the force. Each + and − symbol represents the same quantity of charge.

    a.

    b.

    c.

    d.

23. For each group of charges, use a **black** pen or pencil to draw the forces acting on the gray positive charge. Then use a **red** pen or pencil to show the net force on the gray charge. Label $\vec{F}_{net}$.

    a.

    b.

    c.

    d.

    e.

    f.

24. Can you assign charges (positive or negative) so that these forces are correct? If so, show the charges on the figure. (There may be more than one correct response.) If not, why not?

    a.

    b.

    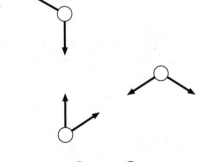

    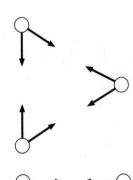

    c.

    d.

25. a. Draw a + on the figure below to show the position or positions where a proton would experience no net force.

|     |     |     (+++)   |     |   (+)   |     |     |

b. Would the force on an electron at this position (or positions) be to the left, to the right, or zero?

26. Draw a – on the figure below to show the position or positions where an electron would experience no net force.

|     |     |     (+++)   |     |   (–)   |     |     |

27. The gray positive charge experiences a net force due to two other charges: the +1 charge that is seen and a +4 charge that is not seen. Add the +4 charge to the figure at the correct position.

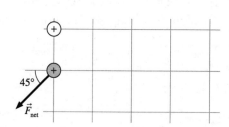

28. In your own words, describe what is meant by a "point charge."

## 25.5  The Concept of a Field

29. This is a uniform gravitational field near the earth's surface. Rank in order, from largest to smallest, the accelerations $a_1$ to $a_3$ of a small mass at points 1, 2, and 3.

    Order:

    Explanation:

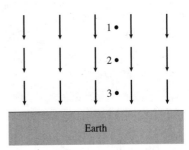

30. This is the gravitational field of the earth. Rank in order, from largest to smallest, the accelerations $a_1$ to $a_3$ of a small mass at points 1, 2, and 3.

    Order:

    Explanation:

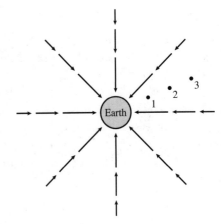

## 25.6  The Field Model

31. At points 1 to 4, draw an electric field vector with the proper direction and whose length is proportional to the electric field strength at that point.

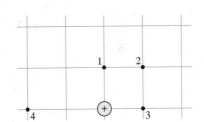

32. The dots are three points in space. The electric field $\vec{E}_1$ at point 1 is shown.

    a. Can you determine the direction of the electric field at point 2? If so, what is it? If not, why not?

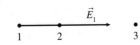

    b. Can you determine the direction of the electric field at point 3? If so, what is it? If not, why not?

33. a. The electric field of a point charge is shown at *one* point in space.

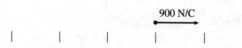

Can you tell if the charge is + or –? If not, why not?

b. Here the electric field of a point charge is shown at two positions in space.

Now can you tell if the charge is + or –? Explain.

c. Can you determine the location of the charge? If so, draw it on the figure. If not, why not?

34. At the three points in space indicated with dots, draw the unit vector $\hat{r}$ that you would use to determine the electric field of the point charge.

a.                                              b.

35. a. This is the unit vector $\hat{r}$ associated with a positive point charge. Draw the electric field vector at this point in space.

b. This is the unit vector $\hat{r}$ associated with a negative point charge. Draw the electric field vector at this point in space.

# 26 The Electric Field

## 26.1 Electric Field Models

## 26.2 The Field of Multiple Point Charges

1. You've been assigned the task of determining the magnitude and direction of the electric field at a point in space. Give a step-by-step procedure of how you will do so. List any objects you will use, any measurements you will make, and any calculations you will need to perform. Make sure that your measurements do not disturb the charges that are creating the field.

2. Is there an electric field at the position of the dot? If so, draw the electric field vector on the figure. If not, what would you need to do to create an electric field at this point?

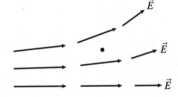

3. This is the electric field in a region of space.
   a. Explain the information that is portrayed in this diagram.

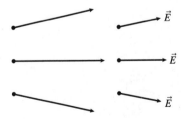

   b. If field vectors were drawn at the same six points but each was only half as long, would the picture represent the same electric field or a different electric field? Explain.

4. Each figure shows two vectors. Can a point charge create an electric field that looks like this at these two points? If so, add the charge to the figure. If not, why not?

**Note:** The dots are the points to which the vectors are attached. There are no charges at these points.

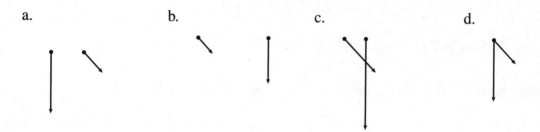

a.     b.     c.     d.

5. At each of the dots, use a **black** pen or pencil to draw and label the electric fields $\vec{E}_1$ and $\vec{E}_2$ due to the two point charges. Make sure that the *relative* lengths of your vectors indicate the strength of each electric field. Then use a **red** pen or pencil to draw and label the net electric field $\vec{E}_{net}$.

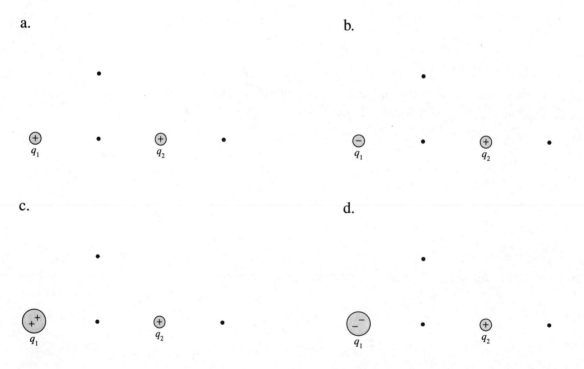

a.                                                                 b.

c.                                                                 d.

6. For each of the figures, use dots to mark any point or points (other than infinity) where $\vec{E} = \vec{0}$.

a.

b.

7. Each figure shows two points near to two charges. Compare the electric field strengths $E_1$ and $E_2$ at these two points. Is $E_1 > E_2$, $E_1 = E_2$, or $E_1 < E_2$?

a.

b.

c.

d.

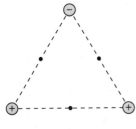

e.

f.

8. For each figure, draw and label the net electric field vector $\vec{E}_{net}$ at each of the points marked with a dot or, if appropriate, label the dot $\vec{E}_{net} = \vec{0}$. The lengths of your vectors should indicate the magnitude of $\vec{E}$ at each point.

a.

b.

c.

d.

9. At the position of the dot, draw fields $\vec{E}_1$ and $\vec{E}_2$ due to $q_1$ and $q_2$, and the net electric field $\vec{E}_{net}$. Then, in the blanks, state whether the x- and y-components of $\vec{E}_{net}$ are positive or negative.

a.

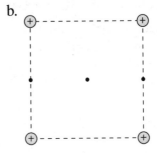

$(E_{net})_x$: _____

$(E_{net})_y$: _____

b.

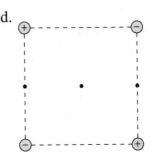

$(E_{net})_x$: _____

$(E_{net})_y$: _____

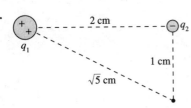
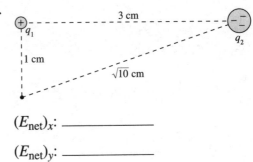

10. Use a **black** pen or pencil to draw the two electric fields $\vec{E}_1$ and $\vec{E}_2$ at each dot. Then use a **red** pen or pencil to draw $\vec{E}_{net}$. The lengths of your vectors should indicate the magnitude of $\vec{E}$ at each point.

a.

b.

·      ·      ·                 ·      ·      ·

·      ·      ·                 ·      ·      ·

⊕    ·    ⊕    ·         ⊖    ·    ⊕    ·
$q_1$       $q_2$           $q_1$       $q_2$

11. Draw the electric field vector at the three points marked with a dot.

    Hint: Think of the charges as horizontal positive/negative pairs, then use superposition.

⊕                 ⊖

⊕       ·       ⊖

⊕                 ⊖

⊕       ·       ⊖

⊕                 ⊖

⊕       ·       ⊖

⊕                 ⊖

12. The figure shows the electic field lines in a region of space. Draw the electric field vectors at the three dots.

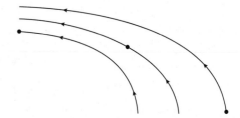

13. The figure shows the electic field lines in a region of space. Rank in order, from largest to smallest, the electric field strengths $E_1$ to $E_4$ at points 1 to 4.

    Order:

    Explanation:

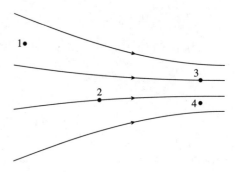

## 26.3  The Electric Field of a Continuous Charge Distribution

14. A small segment of wire contains 10 nC of charge.

    a. The segment is shrunk to one-third of its original length. What is the ratio $\lambda_f/\lambda_i$, where $\lambda_i$ and $\lambda_f$ are the initial and final linear charge densities?

    b. A proton is very far from the wire. What is the ratio $F_f/F_i$ of the electric force on the proton after the segment is shrunk to the force on the proton before the segment is shrunk?

    c. Suppose the original segment of wire is stretched to 10 times its original length. How much charge must be added to the wire to keep the linear charge density unchanged?

15. A wire has initial linear charge density $\lambda_i$. The wire is stretched in length by 50%, and one-third of the charge is removed. What is the ratio $\lambda_f/\lambda_i$, where $\lambda_f$ is the final linear charge density?

16. The figure shows a uniformly charged positive wire. Five small, equally-spaced segments of charge are shown. Use these five segments to estimate the wire's electric field—both magnitude and direction—at each point in space marked with a dot. Draw each $\vec{E}$ on the figure.

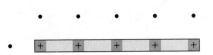

17. Equal-length, equally charged positive and negative wires are placed end-to-end. Draw the electric field at each of the dots.

    Hint: Think about the superposition of the fields of a positive and a negative wire.

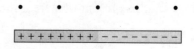

18. Two rings of charge face each other. The total charge on each ring is indicated beneath it. Draw the electric field vector on the axis of the rings at the midpoint between them (at the dot), or label the point $\vec{E} = \vec{0}$.

a.

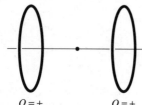

$Q = +$          $Q = +$

b.

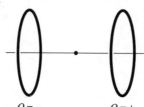

$Q = -$          $Q = +$

c.

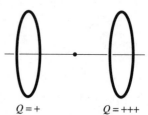

$Q = +$          $Q = +++$

19. The figure shows two charged rods bent into a semicircle. For each, draw the electric field vector at the "center" of the semicircle.

a.

b.

20. A hollow soda straw is uniformly charged. What is the electric field at the center (inside) of the straw? Explain.

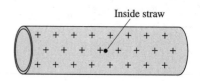

Inside straw

21. An electron experiences a force of magnitude $F$ when it is 1 cm from a very long charged wire with linear charge density $\lambda$. If the charge density is doubled, at what distance from the wire will a proton experience a force of the same magnitude $F$?

## 26.4 The Electric Field of Rings, Planes, and Spheres

22. An irregularly-shaped area of charge has surface charge density $\eta_i$.
    Each dimension ($x$ and $y$) of the area is reduced by a factor of 3.16.

    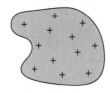

    a. What is the ratio $\eta_f/\eta_i$, where $\eta_f$ is the final surface charge density?

    b. Compare the final force on a electron very far away to the initial force on the same electron.

23. A circular disk has surface charge density 8 nC/cm². What will be the surface charge density if the radius of the disk is doubled?

24. Rank in order, from largest to smallest, the surface charge densities $\eta_1$ to $\eta_4$ of surfaces 1 to 4.

    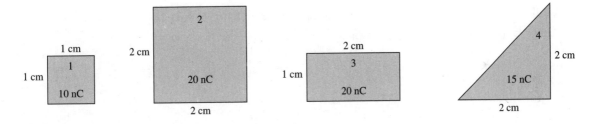

    Order:

    Explanation:

25. The figure shows an edge view of a plane of negative charge. Draw the electric field diagram.

26. A sphere of radius $R_i$ has charge $Q_i$. What happens to the electric field strength at $r = 2R_i$ if:

    a. The quantity of charge is halved?

    b. The radius is halved?

27. A pendulum is made with a ball of mass $m$ and positive charge $q$. It is suspended from a large, uniformly charged positive plate. Gravity is not negligible.

    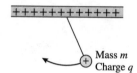

    a. Draw a free-body diagram of the ball when the string is at an angle of 45°.

    b. Would discharging the ball cause the tension in the string to increase, decrease, or stay the same? Explain.

    c. Would discharging the ball cause the period of the pendulum to increase, decrease, or stay the same? Explain.

## 26.5 The Parallel-Plate Capacitor

28. Rank in order, from largest to smallest, the electric field strengths $E_1$ to $E_5$ at each of these points.

    Order:

    Explanation:

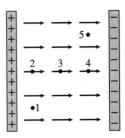

29. A parallel-plate capacitor is constructed of two square plates, size $L{\times}L$, separated by distance $d$. The plates are given charge $\pm Q$. What is the ratio $E_f/E_i$ of the final electric field strength $E_f$ to the initial electric field strength $E_i$ if:

    a. $Q$ is doubled?

    b. $L$ is doubled?

    c. $d$ is doubled?

30. A ball hangs from a thread between two vertical capacitor plates. Initially the ball is neutral. Before the capacitor is charged, the ball hangs exactly in the center.

    a. After the capacitor is charged, is the ball's equilibrium position to the right, to the left, or in the center? Explain.

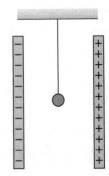

    b. A charged rod is used to give the ball a negative charge. Afterward, is the ball's equilibrium position to the right, to the left, or in the center? Explain.

c. Draw a free-body diagram of the negatively charged ball in static equilibrium.

31. A neutral metal rod is suspended in the center of a parallel-plate capacitor. Then the capacitor is charged as shown.

a. Is the rod now positive, negative, or neutral? Explain.

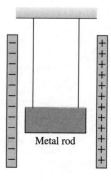

b. Is the rod polarized? If so, draw plusses and minuses on the figure to show the charge distribution. If not, why not?

c. Does the rod swing toward one of the plates, or does it remain in the center? If it swings, which way? Explain.

# 26.6 Motion of a Charged Particle in an Electric Field

# 26.7 Motion of a Dipole in an Electric Field

32. A small positive charge $q$ experiences a force of magnitude $F_1$ when placed at point 1. In terms of $F_1$:

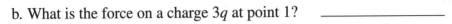

a. What is the force on charge $q$ at point 3?  _____

b. What is the force on a charge $3q$ at point 1?  _____

c. What is the force on a charge $2q$ at point 2?  _____

d. What is the force on a charge $-2q$ at point 2?  _____

33. A small object is released in the center of the capacitor. For each situation, does the object move to the right, to the left, or remain in place? If it moves, does it accelerate or move at constant speed?

a. Positive object released from rest.

b. Negative object released from rest.

c. Neutral object released from rest.

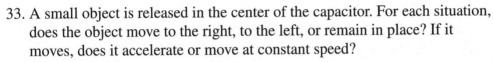

34. Positively and negatively charged objects, with equal masses and equal quantities of charge, enter the capacitor in the directions shown.

a. Use solid lines to draw their trajectories on the figure if their initial velocities are fast.

b. Use dotted lines to draw their trajectories on the figure if their initial velocities are slow.

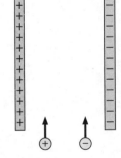

35. An electron is launched from the positive plate at a 45° angle. It does not have sufficient speed to make it to the negative plate. Draw its trajectory on the figure.

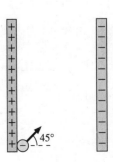

36. A proton and an electron are released from rest in the center of a capacitor.
    a. Compare the forces on the two charges. Are they equal, or is one larger? Explain.

    b. Compare the accelerations of the two charges. Are they equal, or is one larger? Explain.

37. The figure shows an electron orbiting a proton in a hydrogen atom.
    a. What force or forces act on the electron?

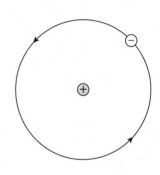

    b. Draw and label the following vectors on the figure: the electron's velocity $\vec{v}$ and acceleration $\vec{a}$, the net force $\vec{F}_{net}$ on the electron, and the electric field $\vec{E}$ at the position of the electron.

38. Does a charged particle always move in the direction of the electric field? If so, explain why. If not, give an example that is otherwise.

39. Three charges are placed at the corners of a triangle. The ++ charge has twice the quantity of charge of the two − charges; the net charge is zero.
    a. Draw the force vectors on each of the charges.
    b. Is the triangle in equilibrium? If not, draw the equilibrium orientation directly beneath the triangle that is shown.
    c. In equilibrium, will the triangle move to the right, move to the left, or remain in place? Explain.

# 27 Gauss's Law

## 27.1 Symmetry

1. An infinite plane of charge is seen edge on. The sign of the charge is not given. Do the electric fields shown below have the same symmetry as the charge? If not, why not?

a.

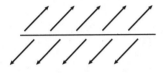

b.

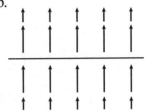

c.

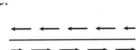

d.

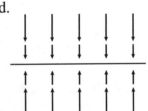

2. Suppose you had a uniformly charged cube. Can you use symmetry alone to deduce the shape of the cube's electric field? If so, sketch and describe the field shape. If not, why not?

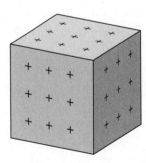

## 27.2 The Concept of Flux

3. The figures shown below are cross sections of three-dimensional closed surfaces. They have a flat top and bottom surface above and below the plane of the page. However, the electric field is everywhere parallel to the page, so there is no flux through the top or bottom surface. The electric field is uniform over each face of the surface. The field strength, in N/C, is shown.

   For each, does the surface enclose a net positive charge, a net negative charge, or no net charge?

a.

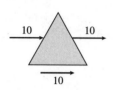

$Q_{net}$ = _____

b.

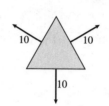

$Q_{net}$ = _____

c.

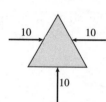

$Q_{net}$ = _____

d.

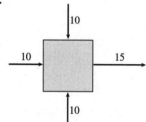

$Q_{net}$ = _____

e.

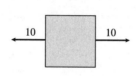

$Q_{net}$ = _____

f.

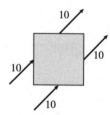

$Q_{net}$ = _____

4. The figures shown below are cross sections of three-dimensional closed surfaces. They have a flat top and bottom surface above and below the plane of the page, but there is no flux through the top or bottom surface. The electric field is uniform over each face of the surface. The field strength, in N/C, is shown.

   Each surface contains no net charge. Draw the missing electric field vector (or write $\vec{E} = \vec{0}$) in the proper direction. Write the field strength beside it.

a.

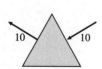

b.

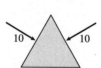

c.

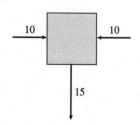

d.

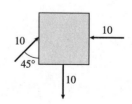

## 27.3 Calculating Electric Flux

5. Draw the area vector $\vec{A}$ for each of these surfaces.

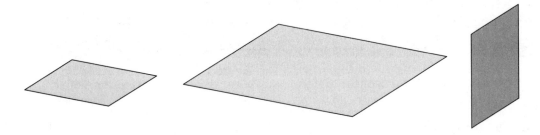

6. How many area vectors are needed to characterize this closed surface?

   Draw them.

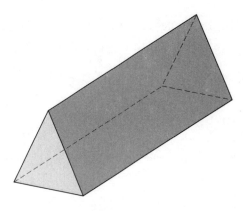

7. The diameter of the circle equals the edge length of the square. Is $\Phi_1$ larger than, smaller than, or equal to $\Phi_2$?

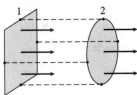

8. Is $\Phi_1$ larger than, smaller than, or equal to $\Phi_2$?

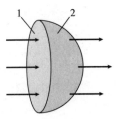

9. A uniform electric field is shown below.

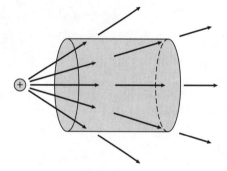

Draw and label an *edge view* of three square surfaces, all the *same size*, for which

a. The flux is maximum.

b. The flux is minimum.

c. The flux has half the value of the flux through square 1.

Draw and label a fourth edge view of a surface with the same orientation as loop 1 but with only 25% as much flux. (Remember, these are *square* surfaces.)

10. Is the net electric flux through each of the closed surfaces below positive, negative, or zero? Explain your reasoning.

a.

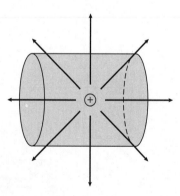

$\Phi =$ _____

b.

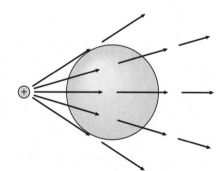

$\Phi =$ _____

c.

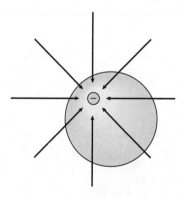

$\Phi =$ _____

d.

$\Phi =$ _____

e.

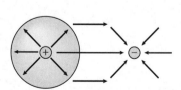

$\Phi =$ _____

f.

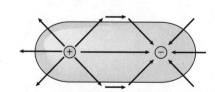

$\Phi =$ _____

## 27.4 Gauss's Law

## 27.5 Using Gauss's Law

11. For each of the closed cylinders shown below, are the electric fluxes through the top, the wall, and the bottom positive (+), negative (−), or zero (0)? Is the net flux positive, negative, or zero?

a.

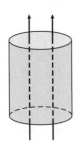

$\Phi_{top} =$ _____

$\Phi_{wall} =$ _____

$\Phi_{bot} =$ _____

$\Phi_{net} =$ _____

b.

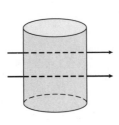

$\Phi_{top} =$ _____

$\Phi_{wall} =$ _____

$\Phi_{bot} =$ _____

$\Phi_{net} =$ _____

c.

$\Phi_{top} =$ _____

$\Phi_{wall} =$ _____

$\Phi_{bot} =$ _____

$\Phi_{net} =$ _____

d.

$\Phi_{top} =$ _____

$\Phi_{wall} =$ _____

$\Phi_{bot} =$ _____

$\Phi_{net} =$ _____

e.

$\Phi_{top} =$ _____

$\Phi_{wall} =$ _____

$\Phi_{bot} =$ _____

$\Phi_{net} =$ _____

f.

$\Phi_{top} =$ _____

$\Phi_{wall} =$ _____

$\Phi_{bot} =$ _____

$\Phi_{net} =$ _____

g.

$\Phi_{top} =$ _____

$\Phi_{wall} =$ _____

$\Phi_{bot} =$ _____

$\Phi_{net} =$ _____

h.

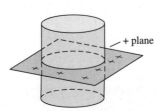

$\Phi_{top} =$ _____

$\Phi_{wall} =$ _____

$\Phi_{bot} =$ _____

$\Phi_{net} =$ _____

i.

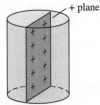

$\Phi_{top} =$ _____

$\Phi_{wall} =$ _____

$\Phi_{bot} =$ _____

$\Phi_{net} =$ _____

12. For this closed cylinder, $\Phi_{top} = -15 \, \text{N}\,\text{m}^2/\text{C}$ and $\Phi_{bot} = 5 \, \text{N}\,\text{m}^2/\text{C}$.
    What is $\Phi_{wall}$?

13. What is the electric flux through each of these surfaces? Give your answer as a multiple of $q/\varepsilon_0$.

a.

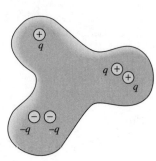

b.

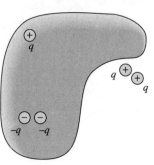

c.

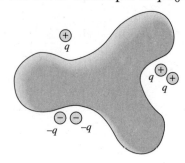

$\Phi_e = $ _____   $\Phi_e = $ _____   $\Phi_e = $ _____

14. What is the electric flux through each of these surfaces?
    Give your answer as a multiple of $q/\varepsilon_0$.

$\Phi_A = $ _____

$\Phi_B = $ _____

$\Phi_C = $ _____

$\Phi_D = $ _____

$\Phi_E = $ _____

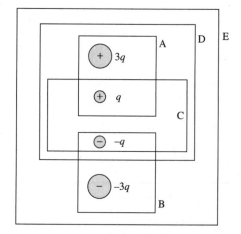

15. A charged balloon expands as it is blown up, increasing in size
    from the initial to final diameters shown. Do the electric fields
    at points 1, 2, and 3 increase, decrease, or stay the same?
    Explain your reasoning for each.

Point 1:

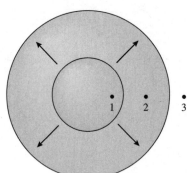

Point 2:

Point 3:

16. Three charges, all the same charge $q$, are surrounded by three spheres of equal radii.

a. Rank in order, from largest to smallest, the fluxes $\Phi_1$, $\Phi_2$, and $\Phi_3$ through the spheres.

Order:

Explanation:

b. Rank in order, from largest to smallest, the electric field strengths $E_1$, $E_2$, and $E_3$ on the surfaces of the spheres.

Order:

Explanation:

17. Two spheres of different diameters surround equal charges. Three students are discussing the situation.

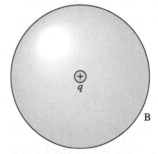

Student 1: The flux through spheres A and B are equal because they enclose equal charges.

Student 2: But the electric field on sphere B is weaker than the electric field on sphere A. The flux depends on the electric field strength, so the flux through A is larger than the flux through B.

Student 3: I thought we learned that flux was about surface area. Sphere B is larger than sphere A, so I think the flux through B is larger than the flux through A.

Which of these students, if any, do you agree with? Explain.

18. A sphere and an ellipsoid surround equal charges. Four students are discussing the situation.

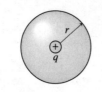   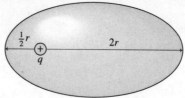

Student 1: The fluxes through A and B are equal because the average radius is the same.

Student 2: I agree that the fluxes are equal, but it's because they enclose equal charges.

Student 3: The electric field is not perpendicular to the surface for B, and that makes the flux through B less than the flux through A.

Student 4: I don't think that Gauss's law even applies to a situation like B, so we can't compare the fluxes through A and B.

Which of these students, if any, do you agree with? Explain.

19. Two parallel, infinite planes of charge have charge densities $2\eta$ and $-\eta$. A Gaussian cylinder with cross section $A$ extends distance $L$ to either side.

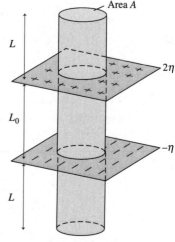

a. Is $\vec{E}$ perpendicular or parallel to the surface at the:

   Top _____   Bottom _____   Wall _____

b. Is the electric field $E_{top}$ emerging from the top surface stronger than, weaker than, or equal in strength to the field $E_{bot}$ emerging from the bottom? Explain.

c. By inspection, write the electric fluxes through the three surfaces in terms of $E_{top}$, $E_{bot}$, $E_{wall}$, $L$, $L_0$, and $A$. (You may not need all of these.)

   $\Phi_{top} =$ _____   $\Phi_{bot} =$ _____   $\Phi_{wall} =$ _____

d. How much charge is enclosed within the cylinder? Write $Q_{in}$ in terms of $\eta$, $L$, $L_0$, and $A$.

   $Q_{in} =$ _____

e. By combining your answers from parts b, c, and d, use Gauss's law to determine the electric field strength above the top plane. Show your work.

# 27.6 Conductors in Electrostatic Equilibrium

20. A small sphere hangs by a thread within a larger, hollow conducting sphere. A charged rod is used to transfer positive charge to the outer surface of the hollow sphere.

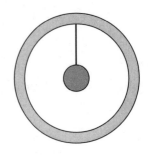

    a. Suppose the thread is an insulator. After the charged rod touches the outer sphere and is removed, are the following surfaces positive, negative, or not charged?

       The small sphere: _____

       The inner surface of the hollow sphere: _____

       The outer surface of the hollow sphere: _____

    b. Suppose the thread is a conductor. After the charged rod touches the outer sphere and is removed, are the following surfaces positive, negative, or not charged?

       The small sphere: _____

       The inner surface of the hollow sphere: _____

       The outer surface of the hollow sphere: _____

21. A small sphere hangs by an insulating thread within a larger, hollow conducting sphere. A conducting wire extends from the small sphere through, but not touching, a small hole in the hollow sphere. A charged rod is used to transfer positive charge to the wire. After the charged rod has touched the wire and been removed, are the following surfaces positive, negative, or not charged?

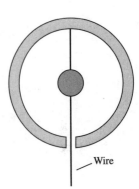

    The small sphere: _____

    The inner surface of the hollow sphere: _____

    The outer surface of the hollow sphere: _____

22. A −10 nC point charge is inside a hole in a conductor. The conductor has no net charge.

    a. What is the total charge on the inside surface of the conductor?

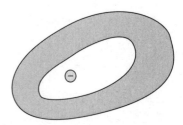

    b. What is the total charge on the outside surface of the conductor?

23. A −10 nC point charge is inside a hole in a conductor. The conductor has a net charge of +10 nC.

    a. What is the total charge on the inside surface of the conductor?

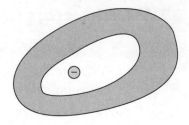

    b. What is the total charge on the outside surface of the conductor?

24. An insulating thread is used to lower a positively charged metal ball into a metal container. Initially, the container has no net charge. Use plus and minus signs to show the charge distribution on the inner and outer surfaces of the container and any charge on the ball. (The ball's charge is already shown in the first frame.)

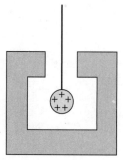

Ball hasn't touched

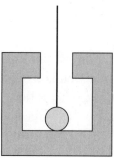

Ball has touched

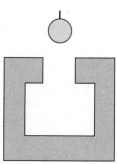

Ball has been withdrawn

# 28 Current and Conductivity

## 28.1 The Electron Current

1. Two wires connect a light bulb to a battery, completing a circuit and causing the bulb to glow. Do the simple observations and measurements that you can make on this circuit prove that something is *flowing* through the wires? If so, state what observations and/or measurements are relevant and the steps by which you can then infer that something must be flowing. If not, can you offer an alternative hypothesis about why the bulb glows that is at least plausible and that could be tested?

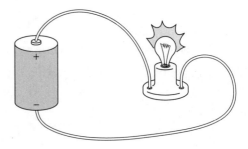

2. Consider a light bulb circuit such as the one in Exercise 1. Are the simple observations and measurements you can make on this circuit able to distinguish a current composed of positive charge carriers from a current composed of negative charge carriers? If so, describe how you can tell which it is. If not, why not?

3. a. Describe an experiment that provides evidence that current consists of charge flowing through a conductor. Use both pictures and words.

b. One model of current is that it consists of the motion of discrete charged particles. Another model is that current is the flow of a continuous charged fluid. Does the experiment you described in part a provide evidence in favor of either one of these models? If so, describe how.

4. Describe experimental evidence to support the claim that the charge carriers in metals are electrons. Use both pictures and words.

5. Are the charge carriers always electrons? If so, why is this the case? If not, describe a situation in which a current is due to some other charge carrier.

# 28.2  Creating a Current

6. What *causes* electrons to move through a wire as a current?

7. The electron drift speed in a wire is exceedingly slow—typically only a fraction of a millimeter per second. Yet when you turn on a light switch, a light bulb several meters away seems to come on instantly. Explain how to resolve this apparent paradox.

8. The figure shows a segment of a current-carrying metal wire.

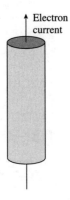

Electron current

   a. Is there an electric field inside the wire? If so, draw and label an arrow on the figure to show its direction. If not, why not?

   b. If there is an electric field, draw on the figure a possible arrangement of charges that could be the source charges causing the field.

9. Is this a possible surface charge distribution for a current-carrying wire?

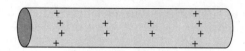

If so, in which direction is the current? If not, why not?

10. a. If the electrons in a current-carrying wire collide with the positive ions *more* frequently, does their drift speed increase or decrease? Explain.

   b. Does an increase in the collision frequency make the wire a better conductor or a worse conductor? Explain.

# 28.3 Batteries

No exercises for this section.

## 28.4 Current and Current Density

11. What is the difference between current and current density?

12. The figure shows a segment of a current-carrying metal wire.

    a. Draw an arrow on the figure, using a **black** pen or pencil, to show the direction of motion of the charge carriers.

    b. Draw an arrow on the figure, using a **red** pen or pencil, to show the direction of the electric field.

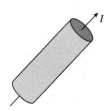

13. Is $I_2$ greater than, less than, or equal to $I_1$? Explain.

14. All wires in this figure are made of the same material and have the same diameter. Rank in order, from largest to smallest, the currents $I_1$ to $I_4$.

    Order:

    Explanation:

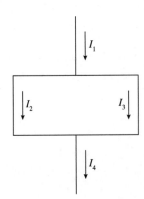

15. A light bulb is connected to a battery with 1-mm-diameter wires. The bulb is glowing.

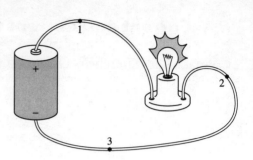

a. Draw arrows at points 1, 2, and 3 to show the direction of the electric field at those points. (The points are *inside* the wire.)

b. Rank in order, from largest to smallest, the field strengths $E_1$, $E_2$, and $E_3$.

Order:

Explanation:

16. A wire carries a 4 A current. What is the current in a second wire that delivers twice as much charge in half the time?

## 28.5 Conductivity and Resistivity

17. Metal 1 and metal 2 are each formed into 1-mm-diameter wires. The electric field needed to cause a 1 A current in metal 1 is larger than the electric field needed to cause a 1 A current in metal 2. Which metal has the larger conductivity? Explain.

18. If a metal is heated, does its conductivity increase, decrease, or stay the same? Explain.

19. Wire 1 has twice the diameter and half the electric field of wire 2. What is the ratio $I_1/I_2$?

20. Wire 1 and wire 2 are made from the same metal. Wire 2 has a larger diameter than wire 1. The electric field strengths $E_1$ and $E_2$ are equal.

   a. Compare the values of the two current densities. Is $J_1$ greater than, less than, or equal to $J_2$? Explain.

   b. Compare the values of the currents $I_1$ and $I_2$.

   c. Compare the values of the electron drift speeds $(v_d)_1$ and $(v_d)_2$.

21. A wire consist of two segments of different diameters but made from the same metal. The current in segment 1 is $I_1$.

   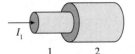

   a. Compare the values of the currents in the two segments. Is $I_2$ greater than, less than, or equal to $I_1$? Explain.

   b. Compare the values of the current densities $J_1$ and $J_2$.

   c. Compare the strengths of the electric fields $E_1$ and $E_2$ in the two segments.

   d. Compare the values of the electron drift speeds $(v_d)_1$ and $(v_d)_2$.

22. A wire consist of two equal-diameter segments. Their conductivities differ, with $\sigma_2 > \sigma_1$. The current in segment 1 is $I_1$.

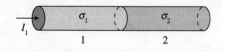

a. Compare the values of the currents in the two segments. Is $I_2$ greater than, less than, or equal to $I_1$? Explain.

b. Compare the strengths of the current densities $J_1$ and $J_2$.

c. Compare the strengths of the electric fields $E_1$ and $E_2$ in the two segments.

d. Compare the values of the electron drift speeds $(v_d)_1$ and $(v_d)_2$.

# 29 The Electric Potential

## 29.1 Electric Potential Energy

## 29.2 The Potential Energy of Point Charges

1. Positive and negative point charges are inside a parallel-plate capacitor. The point charges interact only with the capacitor, not with each other.

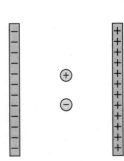

a. Use a **black** pen or pencil to draw the electric field vectors inside the capacitor.

b. Use a **red** pen or pencil to draw the forces acting on the two charges.

c. Pick a point of your choosing for the zero of potential energy. Label it "$U = 0$" on the diagram.

d. Is the potential energy of the *positive* point charge positive, negative, or zero? Explain.

e. In which direction (right, left, up, or down) does the potential energy of the positive charge decrease? Explain.

f. In which direction will the positive charge move if released from rest? Use the concept of energy to explain your answer.

g. Does your answer to part f agree with the force vector you drew in part b? _____

h. Repeat steps d to g for the *negative* point charge.

2. Charge $q$ is fired through a small hole in the positive plate of a capacitor.

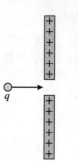

   a. If $q$ is a positive charge, does it speed up or slow down inside the capacitor? Answer this question twice:

      i.   Using the concept of force.

      ii.  Using the concept of energy.

   b. Repeat part a for $q$ as a negative charge.

3. Charge $q$ is fired toward a stationary positive point charge.

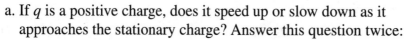  

   a. If $q$ is a positive charge, does it speed up or slow down as it approaches the stationary charge? Answer this question twice:

      i.   Using the concept of force.

      ii.  Using the concept of energy.

   b. Repeat part a for $q$ as a negative charge.

4. a. Charge $q_1 = 3$ nC is distance $r$ from a positive point charge $Q$. Charge $q_2 = 1$ nC is distance $2r$ from $Q$. What is the ratio $U_1/U_2$ of their potential energies due to their interactions with $Q$?

   b. Charge $q_1 = 3$ nC is distance $d$ from the negative plate of a parallel-plate capacitor. Charge $q_2 = 1$ nC is distance $2d$ from the negative plate. What is the ratio $U_1/U_2$ of their potential energies?

5. *Why* is the potential energy of two opposite charges a negative number? (Note: Saying that the formula gives a negative number is not an explanation.)

6. The figure shows the potential energy of a positively charged particle in a region of space.

   a. What arrangement of source charges is responsible for this potential energy? Draw the source above the axis below.

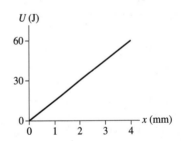

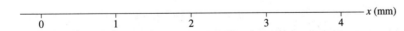

   b. With what kinetic energy should the charged particle be launched from $x = 0$ mm to have a turning point at $x = 3$ mm? Explain.

c. How much kinetic energy does the charged particle of part b have as it passes $x = 2$ mm?

7. The figure shows the potential energy of a proton ($q = +e$) and a lead nucleus ($q = +82e$). The horizontal scale is in units of *femtometers*, where 1 fm = 1 femtometer = $10^{-15}$ m.

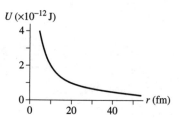

a. A proton is fired toward a lead nucleus from very far away. How much initial kinetic energy does the proton need to reach a turning point 10 fm from the nucleus? Explain.

b. How much kinetic energy does the proton of part a have when it is 20 fm from the nucleus and moving toward it, before the collision?

c. How much kinetic energy does the proton of part a have when it is 20 fm from the nucleus and moving away from it, after the collision?

8. An electron ($q = -e$) completes half of a circular orbit of radius $r$ around a nucleus with $Q = +3e$.

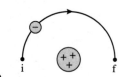

a. How much work is done on the electron as it moves from i to f? Give either a numerical value or an expression from which you could calculate the value if you knew the radius. Justify your answer.

b. By how much does the electric potential energy change as the electron moves from i to f?

c. Is the electron's speed at f greater than, less than, or equal to its speed at i?

d. Are your answers to parts a and c consistent with each other?

9. An electron moves along the trajectory from i to f.
   a. Does the electric potential energy increase, decrease, or stay the
      same? Explain.

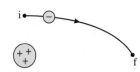

   b. Is the electron's speed at f greater than, less than, or equal to its speed at i? Explain.

10. Inside a parallel-plate capacitor, two protons are launched with the same speed from point 1. One proton moves along the path from 1 to 2, the other from 1 to 3. Points 2 and 3 are the same distance from the positive plate.

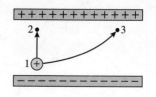

a. Is $\Delta U_{1 \to 2}$, the change in potential energy along the path $1 \to 2$, larger than, smaller than, or equal to $\Delta U_{1 \to 3}$? Explain.

b. Is the proton's speed $v_2$ at point 2 larger than, smaller than, or equal to $v_3$? Explain.

## 29.3  The Potential Energy of a Dipole

11. Rank in order, from most positive to most negative, the potential energies $U_1$ to $U_6$ of these six electric dipoles in a uniform electric field.

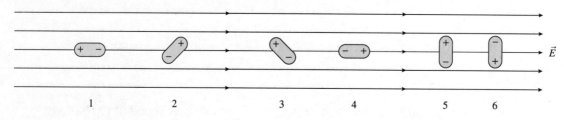

Order:

Explanation:

# 29.4 The Electric Potential

12. Charged particles with $q = +0.1$ C are fired with 10 J of kinetic energy toward a region of space in which there is an electric potential. The figure shows the kinetic energy of the charged particles as they arrive at nine different points in the region. Determine the electric potential at each of these points. Write the value of the potential above each of the dots. Assume that the particles start from a point where the electric potential is zero.

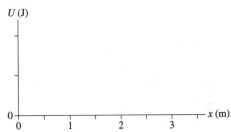

13. a. The graph on the left shows the electric potential along the *x*-axis. Use the axes on the right to draw a graph of the potential energy of a 0.1 C charged particle in this region of space. Provide a numerical scale on the energy axis.

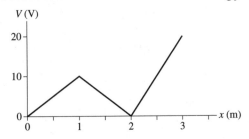

b. If the charged particle is shot toward the right from $x = 1$ m with 1.0 J of kinetic energy, where is its turning point? Explain.

c. Will the charged particle of part b ever reach $x = 0$ m? If so, how much kinetic energy will it have at that point? If not, why not?

## 29.5 The Electric Potential Inside a Parallel-Plate Capacitor

14. Rank in order, from largest to smallest, the electric potentials $V_1$ to $V_5$ at points 1 to 5.

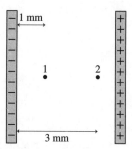

    Order:

    Explanation:

15. The figure shows two points inside a capacitor. Let $V = 0$ V at the negative plate.

    a. What is the ratio $V_2/V_1$ of the electric potential at these two points? Explain.

    b. What is the ratio $E_2/E_1$ of the electric field strength at these two points? Explain.

16. The figure shows two capacitors, each with a 3 mm separation. A proton is released from rest in the center of each capacitor.

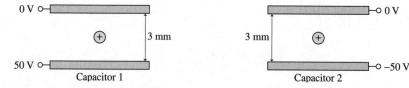

    a. Draw an arrow on each proton to show the direction it moves.

    b. Which proton reaches a capacitor plate first? Or are they simultaneous? Explain.

17. A capacitor with plates separated by distance $d$ is charged to a potential difference $\Delta V_C$. All wires and batteries are disconnected, then the two plates are pulled apart (with insulated handles) to a new separation of distance $2d$.

   a. Does the capacitor charge $Q$ change as the separation increases? If so, by what factor? If not, why not?

   b. Does the electric field strength $E$ change as the separation increases? If so, by what factor? If not, why not?

   c. Does the potential difference $\Delta V_C$ change as the separation increases? If so, by what factor? If not, why not?

18. Each figure shows a contour map on the left and a set of graph axes on the right. Draw a graph of $V$ versus $x$. Your graph should be a straight line or a smooth curve.

a.
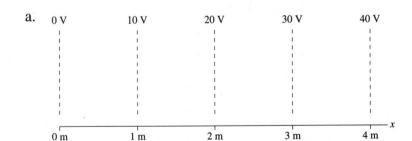
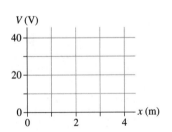

b.
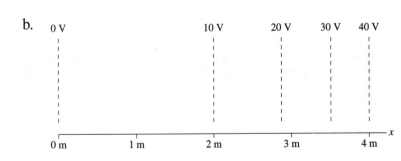
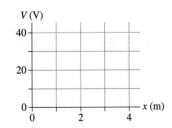

c.
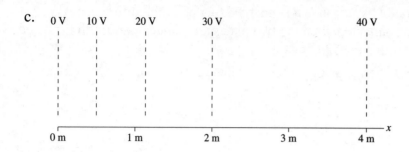

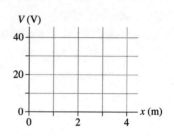

19. Each figure shows a *V*-versus-*x* graph on the left and an *x*-axis on the right. Assume that the potential varies with *x* but not with *y*. Draw a contour map of the electric potential. There should be a uniform difference between equipotential lines, and each equipotential line should be labeled.

a.
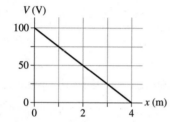

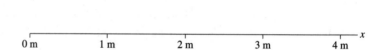

b.
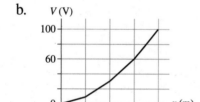

# 29.6  The Electric Potential of a Point Charge

20. Rank in order, from largest to smallest, the electric potentials $V_1$ to $V_5$ at points 1 to 5.

    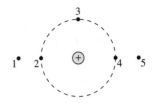

    Order:

    Explanation:

21. Rank in order, from most positive to most negative, the electric potentials $V_1$ to $V_5$ at points 1 to 5.

    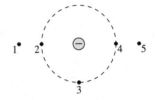

    Order:

    Explanation:

22. The figure shows two points near a positive point charge.

    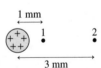

    a. What is the ratio $V_1/V_2$ of the electric potentials at these two points? Explain.

    b. What is the ratio $E_1/E_2$ of the electric field strengths at these two points? Explain.

23. A 1 nC positive point charge is located at point A. The electric potential at point B is

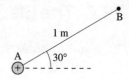

   a. 9 V.           b. 9·sin30° V.     c. 9·cos30° V.      d. 9·tan30° V.

   Explain the reason for your choice.

24. An inflatable metal balloon of radius $R$ is charged to a potential of 1000 V. After all wires and batteries are disconnected, the balloon is inflated to a new radius $2R$.

   a. Does the potential of the balloon change as it is inflated? If so, by what factor? If not, why not?

   b. Does the potential at a point at distance $r = 4R$ change as the balloon is inflated? If so, by what factor? If not, why not?

## 29.7 The Electric Potential of Many Charges

25. Each figure below shows three points in the vicinity of two point charges. The charges have equal magnitudes. Rank in order, from largest to smallest, the potentials $V_1$, $V_2$, and $V_3$.

a.

```
  •      (+)      :      (+)      •
  1               2               3
```

b.

```
  :      (+)      :      (−)      •
  1               2               3
```

c.

```
              1•

   (+)       2•       (+)

              3•
```

d.

```
              1•

   (+)       2•       (−)

              3•
```

26. On the axes below, draw a graph of $V$ versus $x$ for the two point charges shown.

a.

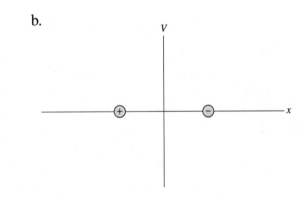

b.

27. For each pair of charges below, are there any points (other than at infinity) at which the electric potential is zero? If so, identify them on the figure with a dot and a label. If not, why not?

a.     ı      ı      ı     (+ +)      ı      ı     (+)      ı      ı      ı

b.     ı      ı      ı     (+ +)      ı      ı     (−)      ı      ı      ı

28. For each pair of charges below, at which grid point or points could a double-negative point charge ($q = -2$) be placed so that the potential at the dot is 0 V? There may be more than one possible point. Draw the charge on the figure at all points that work.

a.

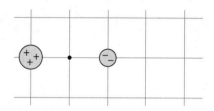

b.

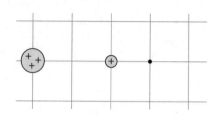

29. The graph shows the electric potential along the $x$-axis due to point charges on the $x$-axis.

   a. Draw the charges *on the axis of the figure*. Note that the charges may have different magnitudes.

   b. An electron is placed at $x = 2$ cm. Is its potential energy positive, negative, or zero? Explain.

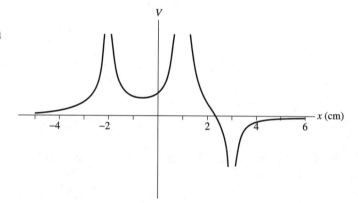

   c. If the electron is released from rest at $x = 2$ cm, will it move right, move left, or remain at $x = 2$ cm? Base your explanation on energy concepts.

30. A ring has radius $R$ and charge $Q$. The ring is shrunk to a new radius $\frac{1}{2}R$ with no change in its charge. By what factor does the on-axis potential at $z = R$ increase?

# 30 Potential and Field

## 30.1 Connecting Potential and Field

1. The top graph shows the *x*-component of $\vec{E}$ as a function of *x*. On the axes below the graph, draw the graph of *V* versus *x* in this same region of space. Let $V = 0$ V at $x = 0$ m. Include an appropriate vertical scale. (Hint: Integration is the area under the curve.)

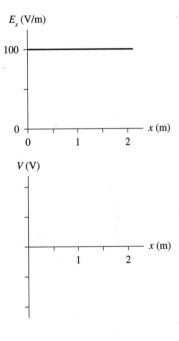

2. a. Suppose that $\vec{E} = \vec{0}$ V/m throughout some region of space. Can you conclude that $V = 0$ V in this region? Explain.

   b. Suppose that $V = 0$ V throughout some region of space. Can you conclude that $\vec{E} = \vec{0}$ V/m in this region? Explain.

## 30.2 Finding the Electric Field from the Potential

3. The top graph on the right shows the electric potential as a function of $x$. On the axes below the graph, draw the graph of $E_x$ versus $x$ in this same region of space.

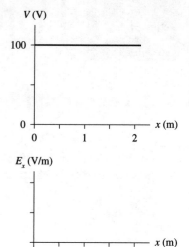

4. For each contour map:

   i.   Estimate the electric fields $\vec{E}_a$ and $\vec{E}_b$ at points a and b. Don't forget that $\vec{E}$ is a vector. Show how you made your estimate.

   ii.  Draw electric field vectors on top of the contour map.

a.

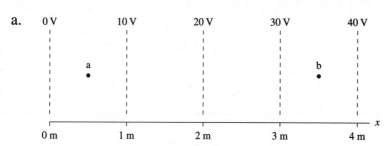

$\vec{E}_a = $ _____

$\vec{E}_b = $ _____

b.

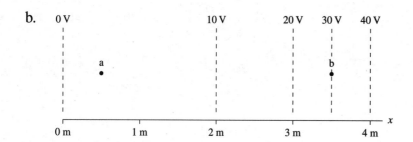

$\vec{E}_a = $ _____

$\vec{E}_b = $ _____

5. The top graph shows $E_x$ versus $x$ for an electric field that is parallel to the $x$-axis.

   a. Draw the graph of $V$ versus $x$ in this region of space. Let $V = 0$ V at $x = 0$ m. Add an appropriate scale on the vertical axis. (Hint: Integration is the area under the curve.)

   b. Draw a contour map above the $x$-axis on the right. Space your equipotential lines every 20 volts and label each equipotential line.

   c. Draw electric field vectors on top of the contour map.

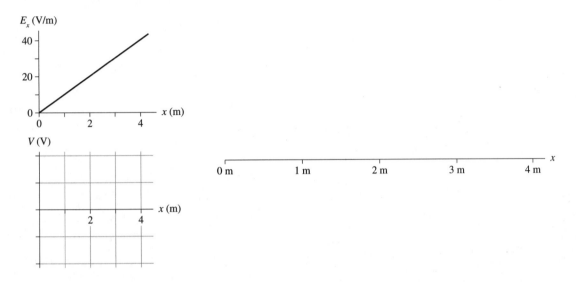

6. Draw the electric field vectors at the dots on this contour map. The length of each vector should be proportional to the field strength at that point.

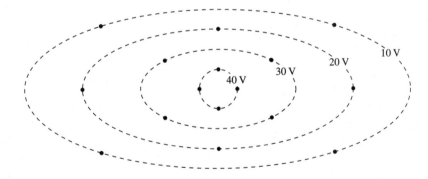

7. Draw the electric field vectors at the dots on this contour map. The length of each vector should be proportional to the field strength at that point.

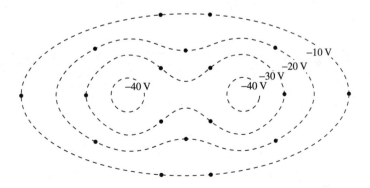

8. Rank in order, from largest to smallest, the electric field strengths $E_1$, $E_2$, $E_3$, and $E_4$ at the four labeled points.

   Order:

   Explanation:

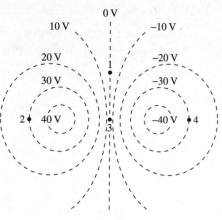

9. For each of the figures below, is this a physically possible potential map if there are no free charges in this region of space? If so, draw an electric field line diagram on top of the potential map. If not, why not?

   a.
   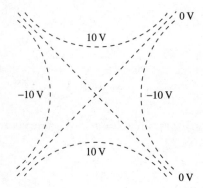

   b.
   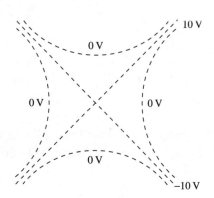

10. The figure shows an electric field diagram. Dotted lines 1 and 2 are two surfaces in space, not physical objects.

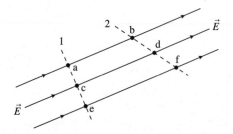

   a. Is the electric potential at point a higher than, lower than, or equal to the electric potential at point b? Explain.

   b. Rank in order, from largest to smallest, the potential differences $\Delta V_{ab}$, $\Delta V_{cd}$, and $\Delta V_{ef}$.

   Order:

   Explanation:

   c. Is surface 1 an equipotential surface? What about surface 2? Explain why or why not.

## 30.3  A Conductor in Electrostatic Equilibrium

11. The figure shows a negatively charged electroscope. The gold leaf stands away from the rigid metal post. Is the electric potential of the leaf higher than, lower than, or equal to the potential of the post? Explain.

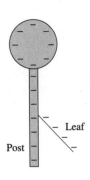

12. Two metal spheres are connected by a metal wire that has a switch in the middle. Initially the switch is open. Sphere 1, with the larger radius, is given a positive charge. Sphere 2, with the smaller radius, is neutral. Then the switch is closed. Afterward, sphere 1 has charge $Q_1$, is at potential $V_1$, and the electric field strength at its surface is $E_1$. The values for sphere 2 are $Q_2$, $V_2$, and $E_2$.

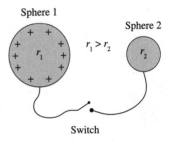

   a. Is $V_1$ larger than, smaller than, or equal to $V_2$? Explain.

   b. Is $Q_1$ larger than, smaller than, or equal to $Q_2$? Explain.

   c. Is $E_1$ larger than, smaller than, or equal to $E_2$? Explain.

13. The figure shows a hollow metal shell. A negatively charged rod touches the top of the sphere, transferring charge to the shell. Then the rod is removed.

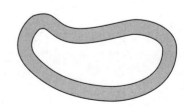

   a. Show on the figure the equilibrium distribution of charge.

   b. Draw the electric field diagram.

14. The figure shows two flat metal electrodes that are held at potentials of 100 V and 0 V.

   a. Sketch a reasonable approximation of the 20 V, 40 V, 60 V, and 80 V equipotential lines.

   b. Draw enough electric field lines to indicate the shape of the electric field.

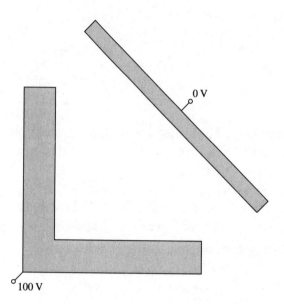

## 30.4  Sources of Potential

## 30.5  Connecting Potential and Current

15. The figure shows two 3 V batteries with metal wires attached to each end. Points a and c are *inside* the wire. Point b is inside the battery. For each figure:
    - What are the potential differences $\Delta V_{12}$, $\Delta V_{23}$, $\Delta V_{34}$, and $\Delta V_{14}$?
    - Does the electric field at a, b, and c point left, right, up, or down? Or is $\vec{E} = \vec{0}$?

a.

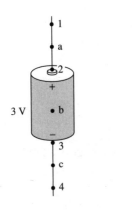

b.

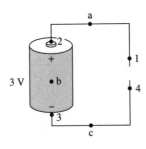

$\Delta V_{12} = $ _____   $\Delta V_{23} = $ _____         $\Delta V_{12} = $ _____   $\Delta V_{23} = $ _____

$\Delta V_{34} = $ _____   $\Delta V_{14} = $ _____         $\Delta V_{34} = $ _____   $\Delta V_{14} = $ _____

$\vec{E}_a$ _____   $\vec{E}_b$ _____   $\vec{E}_c$ _____        $\vec{E}_a$ _____   $\vec{E}_b$ _____   $\vec{E}_c$ _____

16. A continuous metal wire connects the two ends of a 3 V battery with a rectangular loop. The negative terminal of the battery has been chosen as the point where $V = 0$ V.

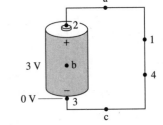

   a. Locate and label the approximate points along the wire where $V = 3$ V, $V = 2$ V, and $V = 1$ V.

   b. Points a and c are *inside* the wire. Point b is inside the battery. Does the electric field at a, b, and c point left, right, up, or down? Or is $\vec{E} = \vec{0}$?

   $\vec{E}_a$ _____   $\vec{E}_b$ _____   $\vec{E}_c$ _____

   c. Estimate the value of $\Delta V_{14}$. Explain how you did so.

   d. In moving through the *wire* from point 2 to point 3, does the potential increase, decrease, or not change? If the potential changes, by how much does it change?

e. In moving through the *battery* from point 2 to point 3, does the potential increase, decrease, or not change? If the potential changes, by how much does it change?

f. In moving all the way around the loop in a clockwise direction, starting from point 2 and ending at point 2, is the net change in the potential positive, negative, or zero?

17. a. Which direction—clockwise or counterclockwise—does an electron travel through the wire? Explain.

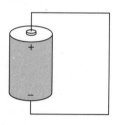

b. Does an electron's electric potential energy increase, decrease, or stay the same as it moves through the wire? Explain.

c. If you answered "decrease" in part b, where does the energy go? If you answered "increase" in part b, where does the energy come from?

d. Which way—up or down—does an electron move through the *battery*? Explain.

e. Does an electron's electric potential energy increase, decrease, or stay the same as it moves through the battery? Explain.

f. If you answered "decrease" in part e, where does the energy go? If you answered "increase" in part e, where does the energy come from?

18. The wires below are all made of the same material. Rank in order, from largest to smallest, the resistances $R_1$ to $R_5$ of these wires.

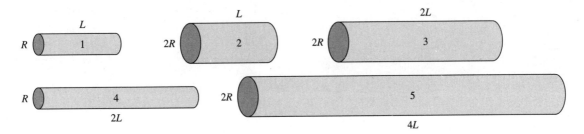

Order:

Explanation:

19. A wire is connected to the terminals of a 6 V battery. What is the potential difference $\Delta V_{ends}$ between the ends of the wire, and what is the current $I$ through the wire, if the wire has the following resistances:

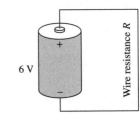

a. $R = 1 \, \Omega$      $\Delta V_{ends} =$ _____     $I =$ _____

b. $R = 2 \, \Omega$      $\Delta V_{ends} =$ _____     $I =$ _____

c. $R = 3 \, \Omega$      $\Delta V_{ends} =$ _____     $I =$ _____

d. $R = 6 \, \Omega$      $\Delta V_{ends} =$ _____     $I =$ _____

20. The two circuits use identical batteries and wires of equal diameters. Rank in order, from largest to smallest, the currents $I_1$, $I_2$, $I_3$, and $I_4$ at points 1 to 4.

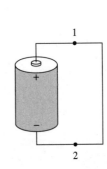
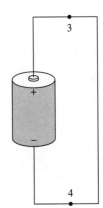

Order:

Explanation:

21. The two circuits use identical batteries and wires of equal diameters. Rank in order, from largest to smallest, the currents $I_1$ to $I_7$ at points 1 to 7.

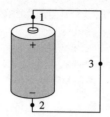 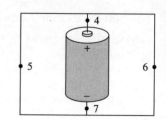

    Order:

    Explanation:

22. Which, if any, of these statements are true? (More than one may be true.)

    i.    A battery supplies the energy to a circuit.

    ii.   A battery is a source of potential difference. The potential difference between the terminals of the battery is always the same.

    iii.  A battery is a source of current. The current leaving the battery is always the same.

    Explain your choice or choices.

## 30.6 Capacitance and Capacitors

## 30.7 The Energy Stored in a Capacitor

23. A parallel-plate capacitor with plate separation $d$ is connected to a battery that has potential difference $\Delta V_{bat}$. Without breaking any of the connections, insulating handles are used to increase the plate separation to $2d$.

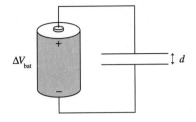

a. Does the potential difference $\Delta V_C$ change as the separation increases? If so, by what factor? If not, why not?

b. Does the capacitance change? If so, by what factor? If not, why not?

c. Does the capacitor charge $Q$ change? If so, by what factor? If not, why not?

d. As the plates are being pulled apart, does current flow cw, ccw, or not at all? Explain.

24. For the capacitor shown, the potential difference $\Delta V_{ab}$ between points a and b is

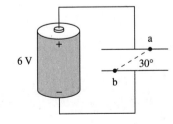

a. 6 V.                b. 6·sin30° V.         c. 6/sin30° V.

d. 6·tan30° V.         e. 6·cos30° V.         f. 6/cos30° V.

Explain your choice.

25. Rank in order, from largest to smallest, the potential differences $(\Delta V_C)_1$ to $(\Delta V_C)_4$ of these four capacitors.

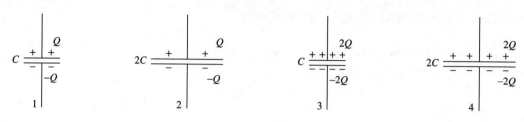

Order:

Explanation:

26. Light bulbs can be used to indicate current flow in a circuit. The brightness of a bulb is proportional to the amount of current passing through it. The figure shows a battery, a switch, two light bulbs, and a capacitor that is initially uncharged.

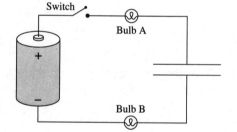

   a. Immediately after the switch is closed, are either or both bulbs glowing? Explain.

   b. If both bulbs are glowing, which is brighter? Or are they equally bright? Explain.

   c. For any bulb (A or B or both) that lights up immediately after the switch is closed, does its brightness increase with time, decrease with time, or remain unchanged? Explain.

27. Each capacitor in the circuits below has capacitance $C$. What is the equivalent capacitance of the group of capacitors?

a.

$C_{eq} =$ _____

b.

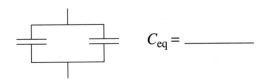

$C_{eq} =$ _____

c.

$C_{eq} =$ _____

d.

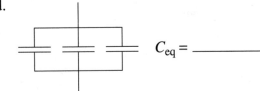

$C_{eq} =$ _____

e.

$C_{eq} =$ _____

f.

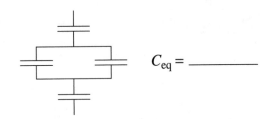

$C_{eq} =$ _____

28. Rank in order, from largest to smallest, the equivalent capacitances $(C_{eq})_1$ to $(C_{eq})_4$ of these four groups of capacitors.

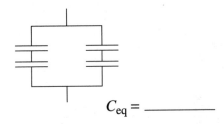

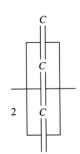

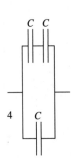

Order:

Explanation:

29. Rank in order, from largest to smallest, the energies $(U_C)_1$ to $(U_C)_4$ stored in each of these capacitors.

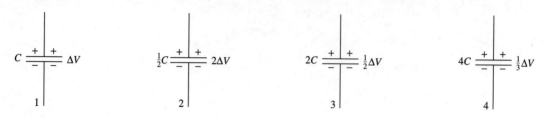

Order:

Explanation:

# 31 Fundamentals of Circuits

## 31.1 Resistors and Ohm's Law

1. The graph shows the current-versus-potential-difference relationship for a resistor $R$.

   a. What is the numerical value of $R$?

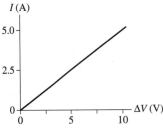

   b. Suppose the length of the resistor is doubled. On the figure, draw the current-versus-potential-difference graph for the longer resistor.

2. For resistors $R_1$ and $R_2$:

   a. Which end (left, right, top, or bottom) is more positive?

   $R_1$ _____    $R_2$ _____

   b. In which direction (such as left to right or top to bottom) does the potential decrease?

   $R_1$ _____

   $R_2$ _____

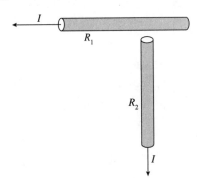

3. Rank in order, from largest to smallest, the currents $I_1$ to $I_4$ through these four resistors.

| + 2 V − | + 1 V − | + 2 V − | + 1 V − |
|---------|---------|---------|---------|
| 2 Ω $I_1$ | 2 Ω $I_2$ | 1 Ω $I_3$ | 1 Ω $I_4$ |

Order:

Explanation:

## 31.2 Circuit Elements and Diagrams

## 31.3 Kirchhoff's Laws and the Basic Circuit

4. The tip of a flashlight bulb is touching the top of a 3 V battery. Does the bulb light? Why or why not?

5. A flashlight bulb is connected to a battery and is glowing. Is current $I_2$ greater than, less than, or equal to current $I_1$? Explain.

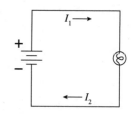

6. Current $I_{in}$ flows into three resistors connected together one after the other. The graph shows the value of the potential as a function of distance.

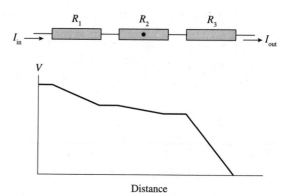

   a. Is $I_{out}$ greater than, less than, or equal to $I_{in}$? Explain.

   b. Rank in order, from largest to smallest, the three resistances $R_1$, $R_2$, and $R_3$.

   Order:

   Explanation:

   c. Is there an electric field at the point inside $R_2$ that is marked with a dot? If so, in which direction does it point? If not, why not?

7. a. In which direction does current flow through resistor $R$?

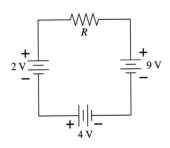

b. Which end of $R$ is more positive? Explain.

c. If this circuit were analyzed in a clockwise direction, what numerical value would you assign to $\Delta V_R$? Why?

d. What value would $\Delta V_R$ have if the circuit were analyzed in a counterclockwise direction?

8. Draw a circuit for which the Kirchhoff loop law equation is
$$6\,V - I \cdot 2\Omega + 3\,V - I \cdot 4\Omega = 0$$
Assume that the analysis is done in a clockwise direction.

9. The wire is broken on the right side of this circuit. What is the potential difference $\Delta V_{12}$ between points 1 and 2? Explain.

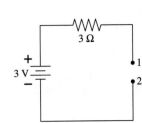

10. The current in a circuit is 2.0 A. The graph shows how the potential changes when going around the circuit in a clockwise direction, starting from the lower left corner. Draw the circuit diagram.

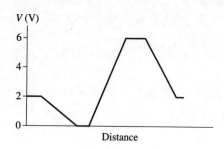

## 31.4 Energy and Power

11. This circuit has two resistors, with $R_1 > R_2$. Which of the two resistors dissipates the larger amount of power? Explain.

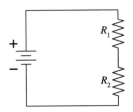

12. Two conductors of equal lengths are connected to a battery by ideal wires. The conductors are made of the same material but have different radii. Which of the two conductors dissipates the larger amount of power? Explain.

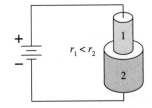

13. Two conductors of equal lengths are connected to a battery by ideal wires. The conductors have the same radii but are made of different materials and have different conductivities $\sigma$. Which of the two conductors dissipates the larger amount of power? Explain.

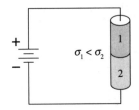

14. A 60 W light bulb and a 100 W light bulb are placed one after the other in a circuit. The battery's emf is large enough that both bulbs are glowing. Which one glows more brightly? Explain.

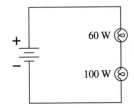

15. Rank in order, from largest to smallest, the powers $P_1$ to $P_4$ dissipated by these four resistors.

Order:

Explanation:

# 31.5 Resistors in Series

# 31.6 Real Batteries

16. What is the equivalent resistance of each group of resistors?

a.

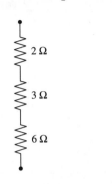

b.

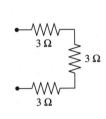

c.

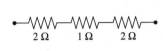

$R_{eq} = $ _____

$R_{eq} = $ _____

$R_{eq} = $ _____

17. The figure shows two circuits. The two batteries are identical and the four resistors all have exactly the same resistance.

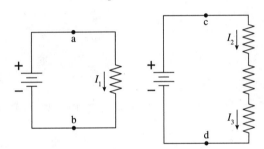

a. Is $\Delta V_{ab}$ larger than, smaller than, or equal to $\Delta V_{cd}$? Explain.

b. Rank in order, from largest to smallest, the currents $I_1$, $I_2$, and $I_3$.

Order:

Explanation:

18. The light bulb in this circuit has a resistance of $1\,\Omega$.

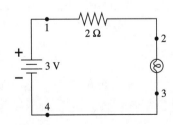

a. What are the values of:

$\Delta V_{12}$ _____

$\Delta V_{23}$ _____

$\Delta V_{34}$ _____

b. Suppose the bulb is now removed from its socket. Then what are the values of:

$\Delta V_{12}$ _____

$\Delta V_{23}$ _____

$\Delta V_{34}$ _____

19. If the value of $R$ is increased, does $\Delta V_{bat}$ increase, decrease, or stay the same? Explain.

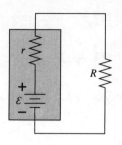

## 31.7 Resistors in Parallel

20. What is the equivalent resistance of each group of resistors?

a.

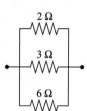

b.

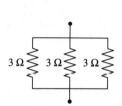

c.

$R_{eq} =$ _____

$R_{eq} =$ _____

$R_{eq} =$ _____

21. a. What fraction of current $I$ goes through the $3\,\Omega$ resistor?

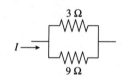

   b. If the $9\,\Omega$ resistor is replaced with a larger resistor, will the fraction of current going through the $3\,\Omega$ resistor increase, decrease, or stay the same?

22. The figure shows five combinations of identical resistors. Rank in order, from largest to smallest, the equivalent resistances $(R_{eq})_1$ to $(R_{eq})_5$.

| 1 | 2 | 3 | 4 | 5 |

Order:

Explanation:

# 31.8  Resistor Circuits

# 31.9  Getting Grounded

23. The circuit shown has a battery and two resistors, with $R_1 > R_2$. Which of the two resistors dissipates the larger amount of power? Explain your reasoning.

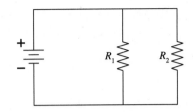

24. Rank in order, from largest to smallest, the three currents $I_1$ to $I_3$.

    Order:

    Explanation:

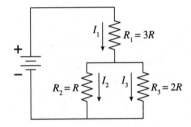

25. The two batteries are identical and the four resistors all have exactly the same resistance.

    a. Compare $\Delta V_{ab}$, $\Delta V_{cd}$, and $\Delta V_{ef}$. Are they all the same? If not, rank them in decreasing order. Explain your reasoning.

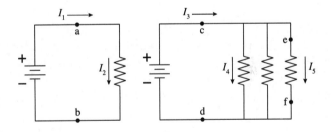

    b. Rank in order, from largest to smallest, the five currents $I_1$ to $I_5$.

    Order:

    Explanation:

**Exercises 26–32:** Assume that all wires are ideal (zero resistance) and that all batteries are ideal (constant potential difference).

26. Initially bulbs A and B are glowing. Then the switch is closed. What happens to each bulb? Does it get brighter, stay the same, get dimmer, or go out? Explain your reasoning.

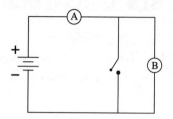

27. a. Bulbs A, B, and C are identical. Rank in order, from most to least, the brightnesses of the three bulbs.

   Order:

   Explanation:

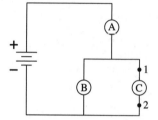

   b. Suppose a wire is connected between points 1 and 2. What happens to each bulb? Does it get brighter, stay the same, get dimmer, or go out? Explain.

28. a. Consider the points a and b. Is the potential difference $\Delta V_{ab} = 0$? If so, why? If not, which point is more positive?

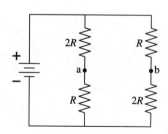

b. If a wire is connected between points a and b, does a current flow through it? If so, in which direction—to the right or to the left? Explain.

29. Bulbs A and B are identical. Initially both are glowing.

a. Bulb A is removed from its socket. What happens to bulb B? Does it get brighter, stay the same, get dimmer, or go out? Explain.

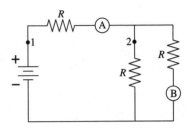

b. Bulb A is replaced. Bulb B is then removed from its socket. What happens to bulb A? Does it get brighter, stay the same, get dimmer, or go out? Explain.

c. The circuit is restored to its initial condition. A wire is then connected between points 1 and 2. What happens to the brightness of each bulb?

30. Initially the light bulb is glowing. It is then removed from its socket.

    a. What happens to the current $I$ when the bulb is removed? Does it increase, stay the same, or decrease? Explain.

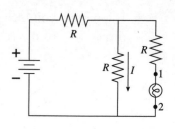

    b. What happens to the potential difference $\Delta V_{12}$ between points 1 and 2? Does it increase, stay the same, decrease, or become zero? Explain.

31. Bulbs A and B are identical and initially both are glowing. Then the switch is closed. What happens to each bulb? Does its brightness increase, stay the same, decrease, or go out? Explain.

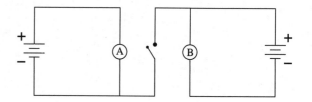

32. Bulbs A and B are identical and initially both are glowing. Then the switch is closed. What happens to each bulb? Does its brightness increase, stay the same, decrease, or go out? Explain.

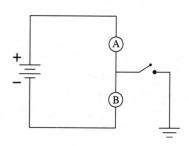

# 31.10 *RC* Circuits

33. The graph shows the voltage as a function of time on a capacitor as it is discharged (separately) through three different resistors. Rank in order, from largest to smallest, the values of the resistances $R_1$ to $R_3$.

    Order:

    Explanation:

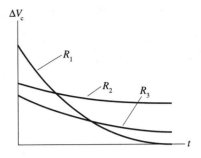

34. The capacitors in each circuit are discharged when the switch closes at $t = 0$ s. Rank in order, from largest to smallest, the time constants $\tau_1$ to $\tau_5$ with which each circuit will discharge.

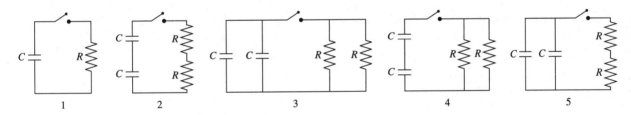

    Order:

    Explanation:

35. The charge on the capacitor is zero when the switch closes at $t = 0$ s.

    a. What will be the current in the circuit after the switch has been closed for a long time? Explain.

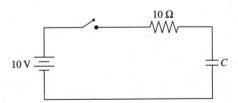

b. Immediately after the switch closes, before the capacitor has had time to charge, the potential difference across the capacitor is zero. What must be the potential difference across the resistor in order to satisfy Kirchhoff's loop law? Explain.

c. Based on your answer to part b, what is the current in the circuit immediately after the switch closes?

d. Sketch a graph of current versus time, starting from just before $t = 0$ s and continuing until the switch has been closed a long time. There are no numerical values for the horizontal axis, so you should think about the *shape* of the graph.

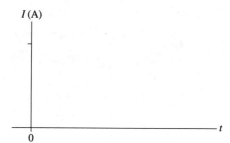

# 32 The Magnetic Field

## 32.1 Magnetism

1. A lightweight glass sphere hangs by a thread. The north pole of a bar magnet is brought near the sphere.

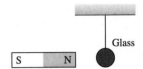

a. Suppose the sphere is electrically neutral. How does it respond?
   i.   It is strongly attracted to the magnet.
   ii.  It is weakly attracted to the magnet.
   iii. It does not respond.
   iv.  It is weakly repelled by the magnet.
   v.   It is strongly repelled by the magnet.
   Explain your choice.

b. How does the sphere respond if it is positively charged? Explain.

2. A metal sphere hangs by a thread. When the north pole of a bar magnet is brought near, the sphere is strongly attracted to the magnet. Then the magnet is reversed and its south pole is brought near the sphere. How does the sphere respond? Explain.

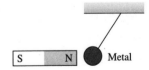

3. The compass needle below is free to rotate in the plane of the page. Either a bar magnet or a charged rod is brought toward the *center* of the compass. Does the compass rotate? If so, in which direction? If not, why not?

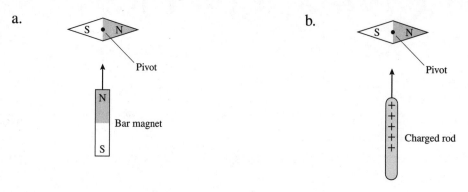

a.

Pivot

Bar magnet

b.

Pivot

Charged rod

4. You have two electrically neutral metal cylinders that exert strong attractive forces on each other. You have no other metal objects. Can you determine if *both* of the cylinders are magnets, or if one is a magnet and the other just a piece of iron? If so, how? If not, why not?

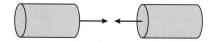

5. Can you think of any kind of object that is repelled by *both* ends of a bar magnet? If so, what? If not, what prevents this from happening?

## 32.2 The Magnetic Field

6. A neutral copper rod, a charged insulator rod, and a bar
   magnet are arranged around a current-carrying wire as
   shown. For each, will it stay where it is? Move toward or
   away from the wire? Rotate clockwise or counterclockwise?
   Explain.

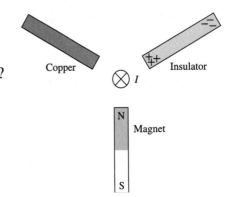

   a. Neutral copper rod:

   b. Insulating rod:

   c. Bar magnet:

7. For each of the current-carrying wires shown, draw a compass needle in its equilibrium
   position at the positions of the dots. Label the poles of the compass needle.

   a.         b.

8. The figure shows a wire directed into the page and a nearby compass needle.
   Is the wire's current going into the page or coming out of the page? Explain.

9. A compass is placed at 12 different positions and its
   orientation is recorded. Use this information to draw
   the magnetic *field lines* in this region of space. Draw
   the field lines on the figure.

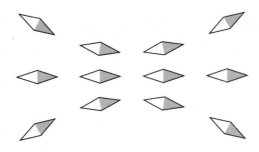

## 32.3 The Source of the Magnetic Field: Moving Charges

10. A positively charged particle moves toward the bottom of the page.

    a. At each of the six number points, show the direction of the magnetic field or, if appropriate, write $\vec{B} = \vec{0}$.

    b. Rank in order, from strongest to weakest, the magnetic field strengths $B_1$ to $B_6$ at these points.

    Order:

    Explanation:

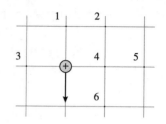

11. The negative charge is moving out of the page, coming toward you. Draw the magnetic field lines in the plane of the page.

12. Two charges are moving as shown. At this instant of time, the net magnetic field at point 2 is $\vec{B}_2 = \vec{0}$.

    a. Is the unlabeled moving charge positive or negative? Explain.

    b. What is the magnetic field direction at point 1? Explain.

    c. What is the magnetic field direction at point 3?

## 32.4 The Magnetic Field of a Current

## 32.5 Magnetic Dipoles

13. The figure shows a current-carrying wire. Draw the magnetic field diagram:

a.

b.

The wire is perpendicular to the page. Draw magnetic field *lines*, then show the magnetic field *vectors* at a few points around the wire.

The wire is in the plane of the page. Show the magnetic field above and below the wire.

14. This current-carrying wire is in the plane of the page. Draw the magnetic field on both sides of the wire.

15. Use an arrow to show the current direction in this wire.

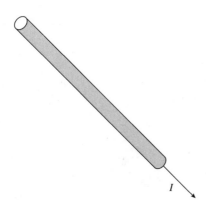

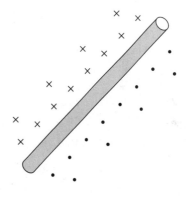

16. Each figure below shows two long straight wires carrying equal currents into and out of the page. At each of the dots, use a **black** pen or pencil to show and label the magnetic fields $\vec{B}_1$ and $\vec{B}_2$ of each wire. Then use a **red** pen or pencil to show the net magnetic field.

a.

Wire 1
⊗

•               •

⊙
Wire 2

b.

Wire 1
⊙

•               •

⊙
Wire 2

17. A long straight wire, perpendicular to the page, passes through a uniform magnetic field. The *net* magnetic field at point 3 is zero.

    a. On the figure, show the direction of the current in the wire.

    b. Points 1 and 2 are the same distance from the wire as point 3, and point 4 is twice as distant. Construct vector diagrams at points 1, 2, and 4 to determine the net magnetic field at each point.

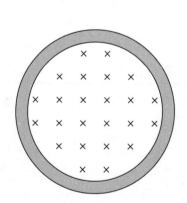

18. A long straight wire passes above one edge of a current loop. Both are perpendicular to the page. $\vec{B}_1 = \vec{0}$ at point 1.

    a. On the figure, show the direction of the current in the loop.

    b. Use a vector diagram to determine the net magnetic field at point 2.

19. The figure shows the magnetic field seen when facing a current loop in the plane of the page.

    a. On the figure, show the direction of the current in the loop.

    b. Is the north pole of this loop at the upper surface of the page or the lower surface of the page? Explain.

20. The current loop exerts a repulsive force on the bar magnet. On the figure, show the direction of the current in the loop. Explain.

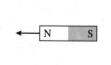

## 32.6 Ampère's Law and Solenoids

21. What is the total current through the area bounded by the closed curve?

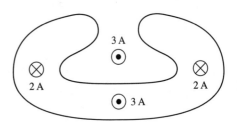

22. The total current through the area bounded by the closed curve is 2 A. What are the size and direction of $I_3$?

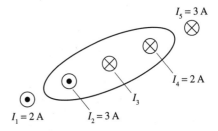

23. The magnetic field above the dotted line is $\vec{B} = (2 \text{ T, right})$. Below the dotted line the field is $\vec{B} = (2 \text{ T, left})$. Each closed loop is 1 m × 1 m. Let's evaluate the line integral of $\vec{B}$ around each of these closed loops by breaking the integration into four steps. We'll go around the loop in a *clockwise* direction. Pay careful attention to signs.

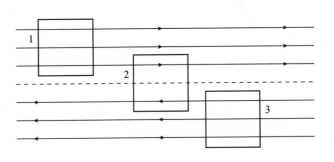

|  | Loop 1 | Loop 2 | Loop 3 |
|---|---|---|---|
| $\int \vec{B} \cdot d\vec{s}$ along left edge | _____ | _____ | _____ |
| $\int \vec{B} \cdot d\vec{s}$ along top | _____ | _____ | _____ |
| $\int \vec{B} \cdot d\vec{s}$ along right edge | _____ | _____ | _____ |
| $\int \vec{B} \cdot d\vec{s}$ along bottom | _____ | _____ | _____ |

The line integral *around* the loop is simply the sum of these four separate integrals:

$\oint \vec{B} \cdot d\vec{s}$ around the loop    _____    _____    _____

24. The strength of a circular magnetic field decreases with increasing radius as shown.

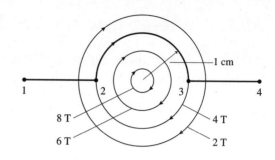

  a. What is $\int_1^2 \vec{B} \cdot d\vec{s}$? _____

   Explain or show your work.

  b. What is $\int_2^3 \vec{B} \cdot d\vec{s}$? _____     Explain or show your work.

  c. What is $\int_3^4 \vec{B} \cdot d\vec{s}$? _____     Explain or show your work.

  d. Combining your answers to parts a to c, what is $\int_1^4 \vec{B} \cdot d\vec{s}$? _____

25. A solenoid with one layer of turns produces the magnetic field strength you need for an experiment when the current in the coil is 3 A. Unfortunately, this amount of current overheats the coil. You've determined that a current of 1 A would be more appropriate. How many additional layers of turns must you add to the solenoid to maintain the magnetic field strength?

26. Rank in order, from largest to smallest, the magnetic fields $B_1$ to $B_3$ produced by these three solenoids.

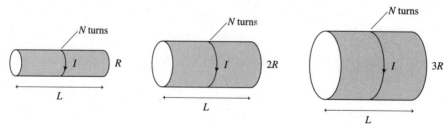

Order:

Explanation:

# 32.7 The Magnetic Force on a Moving Charge

27. For each of the following, draw the magnetic force vector on the charge or, if appropriate, write "$\vec{F}$ into page," "$\vec{F}$ out of page," or "$\vec{F} = \vec{0}$."

a.

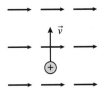

b.

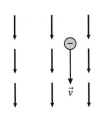

c.

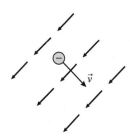

d.

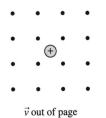

$\vec{v}$ out of page

e.

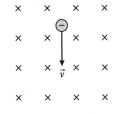

f.

28. For each of the following, determine the sign of the charge (+ or −).

a.

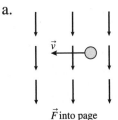

$\vec{F}$ into page

$q = $ \_\_\_\_\_

b.

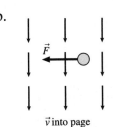

$\vec{v}$ into page

$q = $ \_\_\_\_\_

c.

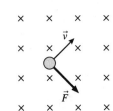

$q = $ \_\_\_\_\_

d.

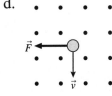

$q = $ \_\_\_\_\_

29. The magnetic field is constant magnitude inside the dotted lines and zero outside. Sketch and label the trajectory of the charge for

a. A very weak field.

b. A moderate field.

c. A strong field.

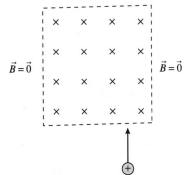

30. A positive ion, initially traveling into the page, is shot through the gap in a magnet. Is the ion deflected up, down, left, or right? Explain.

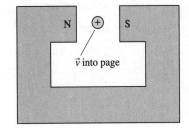

31. A positive ion is shot between the plates of a parallel-plate capacitor.

    a. In what direction is the electric force on the ion?

    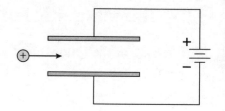

    b. Could a magnetic field exert a magnetic force on the ion that is opposite in direction to the electric force? If so, show the magnetic field on the figure.

32. In a high-energy physics experiment, a neutral particle enters a bubble chamber in which a magnetic field points into the page. The neutral particle undergoes a collision inside the bubble chamber, creating two charged particles. The subsequent trajectories of the charged particles are shown.

    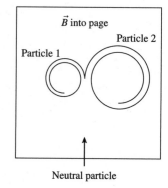

    a. What is the sign (+ or −) of particle 1? _____

       What is the sign (+ or −) of particle 2? _____

    b. Which charged particle leaves the collision with a larger momentum? Explain. (Assume that $|q| = e$ for both particles.)

33. A solenoid is wound as shown and attached to a battery. Two electrons are fired into the solenoid, one from the end and one through a very small hole in the side.

    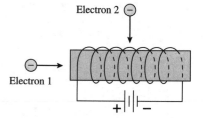

    a. In what direction does the magnetic field inside the solenoid point? Show it on the figure.

    b. Is electron 1 deflected as it moves through the solenoid? If so, in which direction? If not, why not?

    c. Is electron 2 deflected as it moves through the solenoid? If so, in which direction? If not, why not?

34. Two protons are traveling in the directions shown.

    a. Draw the electric forces on each proton.

    b. Draw the magnetic forces on each proton due to the other proton. Explain how you determined the directions.

## 32.8  Magnetic Forces on Current-Carrying Wires

## 32.9  Forces and Torques on Current Loops

35. Three current-carrying wires are perpendicular to the page. Construct a force vector diagram on the figure to find the net force on the upper wire due to the two lower wires.

36. Three current-carrying wires are perpendicular to the page.

   a. Construct a force vector diagram on each wire to determine the direction of the net force on each wire.

   b. Can three *charges* be placed in a triangular pattern so that their force diagram looks like this? If so, draw it. If not, why not?

37. A current-carrying wire passes between two bar magnets. Is there a force on the wire? If so, draw the force vector. If not, why not?

b.

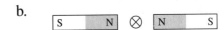

a.

38. A current-carrying wire passes in front of a solenoid that is wound as shown. The wire experiences an upward force. Use arrows to show the direction in which the current enters and leaves the solenoid. Explain your choice.

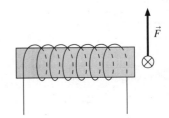

39. Two current-carrying wires cross at right angles.

    a. Draw magnetic force vectors on the wires at the points indicated with dots.

    b. If the wires aren't restrained, how will they behave?

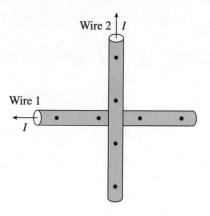

40. A current loop is placed between two bar magnets. Does the loop move to the right, move to the left, rotate clockwise, rotate counterclockwise, some combination of these, or none of these? Explain.

41. A square current loop is placed in a magnetic field as shown.

    a. Does the loop undergo a displacement? If so, is it up, down, left, or right? If not, why not?

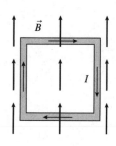

    b. Does the loop rotate? If so, which edge rotates out of the page and which edge into the page? If not, why not?

42. The south pole of a bar magnet is brought toward the current loop. Does the bar magnet attract the loop, repel the loop, or have no effect on the loop? Explain.

## 32.10 Magnetic Properties of Matter

43. A solenoid, wound as shown, is placed next to an unmagnetized piece of iron. Then the switch is closed.

    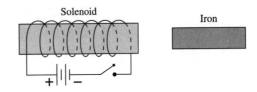

    a. Identify on the figure the north and south poles of the solenoid.

    b. What is the direction of the solenoid's magnetic field as it passes through the iron?

    c. What is the direction of the induced magnetic dipole in the iron?

    d. Identify on the figure the north and south poles of the induced magnetic dipole in the iron.

    e. When the switch is closed, does the iron move left or right? Does it rotate? Explain.

    f. Suppose the iron is replaced by a piece of copper. What happens to the copper when the switch is closed?

# 33 Electromagnetic Induction

## 33.1 Induced Currents

## 33.2 Motional EMF

1. The figures below show one or more metal wires sliding on fixed metal rails in a magnetic field. For each, determine if the induced current flows clockwise, flows counterclockwise, or is zero.

a.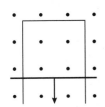

b.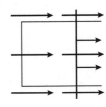

c. 

d.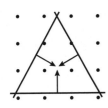

e.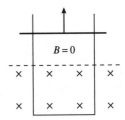

f.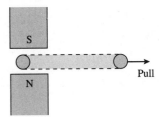

2. A loop of copper wire is being pulled from between two magnetic poles.

   a. Show on the figure the current induced in the loop. Explain your reasoning.

   b. Does either side of the loop experience a magnetic force? If so, draw a vector arrow or arrows on the figure to show any forces.

   c. Label the magnetic poles of the induced current in the loop. Do this on the figure.

   d. Are the magnetic poles you labeled in part c attracted to or repelled by the permanent magnet?

e. Is your answer to part d consistent with your force vectors in part b?

3. You want to insert a loop of copper wire between two permanent magnets. Is there an attractive magnetic force that tends to *pull* the loop in, like a magnet pulls on a paper clip? Or do you need to *push* the loop in against a repulsive force? Give a step-by-step analysis to support your answer.

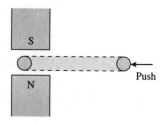

4. A vertical, rectangular loop of copper wire is half in and half out of a horizontal magnetic field. (The field is zero beneath the dotted line.) The loop is released and starts to fall.

a. Add arrows to the figure to show the direction of the induced current in the loop.

b. Is there a net magnetic force on the loop? If so, in which direction? Explain.

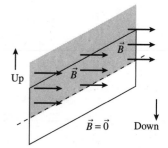

5. Two very thin sheets of copper are pulled through a magnetic field. Do eddy currents flow in the sheet? If so, show them on the figures, with arrows to indicate the direction of flow. If not, why not?

a.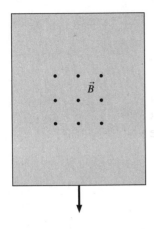

b.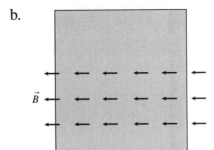

6. The figure shows an edge view of a copper sheet being pulled between two magnetic poles.

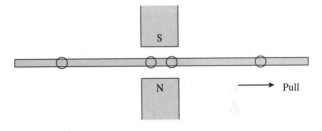

   a. Add a dot or an × to each of the circles to indicate the direction in which eddy currents are flowing in and out of the page.

   b. Label the magnetic poles of any induced current loops.

   c. Do the magnetic poles you labeled in part b experience magnetic forces? If so, add force vectors to the figure to show the directions. If not, why not?

   d. Is there a net magnetic force on the copper sheet? If so, in which direction?

7. An insulating rod pushes a copper loop back and forth. The left edge of the loop, which is always in the magnetic field, oscillates between $x = -L$ and $x = +L$, as shown in the top graph. The right edge of the loop, which includes a lightbulb, is always outside the magnetic field.

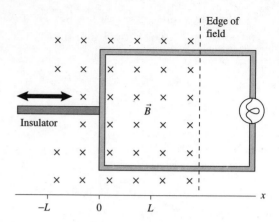

a. Draw the velocity graph for the loop. Make sure it aligns with the position graph above it.

b. Draw a graph of the induced current in the loop as a function of time. Let a clockwise current be a positive number and a counterclockwise current be a negative number.

c. Draw a graph of the brightness of the lightbulb as a function of time.

**Note:** There are no numbers on the vertical scale. The *shape* of each graph is the important result.

# 33.3 Magnetic Flux

8. The figure shows five loops in a magnetic field. The numbers indicate the lengths of the sides and the strength of the field. Rank in order, from largest to smallest, the magnetic fluxes $\Phi_1, \ldots, \Phi_5$. Some may be equal.

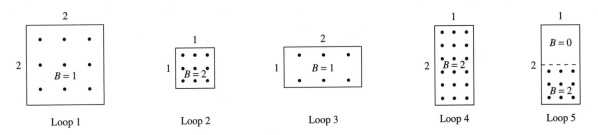

Order:

Explanation:

9. The figure shows four circular loops that are perpendicular to the page. The radius of loops 3 and 4 is twice that of loops 1 and 2. The magnetic field is the same for each. Rank in order, from largest to smallest, the magnetic fluxes $\Phi_1, \ldots, \Phi_4$. Some may be equal.

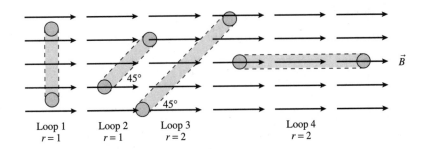

Order:

Explanation:

10. A circular loop rotates at constant speed about an axle through the center of the loop. The figure shows an edge view and defines the angle $\phi$, which increases from 0° to 360° as the loop rotates.

    a. At what angle or angles is the magnetic flux a maximum?

    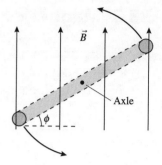

    b. At what angle or angles is the magnetic flux a minimum?

    c. At what angle or angles is the magnetic flux *changing* most rapidly? Explain your choice.

11. A magnetic field is perpendicular to a loop. The graph shows how the magnetic field changes as a function of time, with positive values for $B$ indicating a field into the page and negative values a field out of the page. Several points on the graph are labeled.

    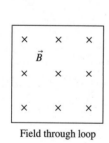

    Field through loop

    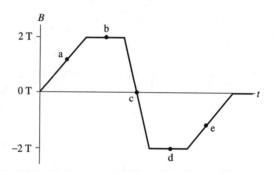

    a. At which lettered point or points is the flux through the loop a maximum?

    b. At which lettered point or points is the flux through the loop a minimum?

    c. At which point or points is the flux changing most rapidly?

    d. At which point or points is the flux changing least rapidly?

## 33.4 Lenz's Law

## 33.5 Faraday's Law

12. Does the loop of wire have a clockwise current, a counterclockwise current, or no current under the following circumstances? Explain.

a. The magnetic field points out of the page and its strength is increasing.

b. The magnetic field points out of the page and its strength is constant.

c. The magnetic field points out of the page and its strength is decreasing.

13. Two loops of wire are stacked vertically, one above the other. Does the upper loop have a clockwise current, a counterclockwise current, or no current at the following times? Explain.

a. Before the switch is closed.

b. Immediately after the switch is closed.

c. Long after the switch is closed.

d. Immediately after the switch is reopened.

14. A loop of wire is perpendicular to a magnetic field. The magnetic field strength as a function of time is given by the top graph. Draw a graph of the current in the loop as a function of time. Let a positive current represent a current that comes out of the top of the loop and enters the bottom of the loop. There are no numbers for the vertical axis, but your graph should have the correct shape and proportions.

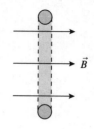

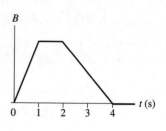

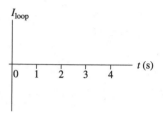

15. A loop of wire is horizontal. A bar magnet is pushed toward the loop from below, along the axis of the loop.

a. What is the current direction in the loop? Explain.

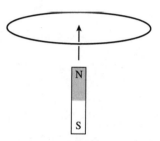

b. Is there a magnetic force on the loop? If so, in which direction? Explain.

Hint: A current loop is a magnetic dipole.

c. Is there a force on the magnet? If so, in which direction?

16. A bar magnet is pushed toward a loop of wire, as shown. Is there a current in the loop? If so, in which direction? If not, why not?

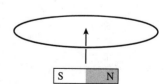

17. A bar magnet is dropped, south pole down, through the center of a loop of wire. The center of the magnet passes the plane of the loop at time $t_c$.

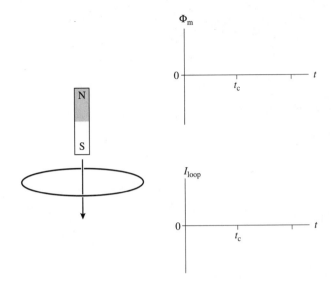

a. Sketch a graph of the magnetic flux through the loop as a function of time.

b. Sketch a graph of the current in the loop as a function of time. Let a clockwise current be a positive number and a counterclockwise current be a negative number.

18. a. As the magnet is inserted into the coil, does current flow right to left or left to right through the current meter? Or is the current zero? Explain.

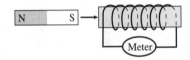

b. As the magnet is held at rest inside the coil, does current flow right to left or left to right through the current meter? Or is the current zero? Explain.

c. As the magnet is withdrawn from the coil, does current flow right to left or left to right through the current meter? Or is the current zero? Explain.

d. If the magnet is inserted into the coil more rapidly than in part a, does the size of the current increase, decrease, or remain the same? Explain.

19. a. Just after the switch on the left coil is closed, does current flow right to left or left to right through the current meter of the right coil? Or is the current zero? Explain.

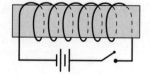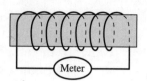

   b. Long after the switch on the left coil is closed, does current flow right to left or left to right through the current meter of the right coil? Or is the current zero? Explain.

   c. Just after the switch on the left coil is reopened, does current flow right to left or left to right through the current meter of the right coil? Or is the current zero? Explain.

20. A solenoid is perpendicular to the page, and its field strength is increasing. Three circular wire loops of equal radii are shown. Rank in order, from largest to smallest, the size of the induced emf in the three rings.

   Order:

   Explanation:

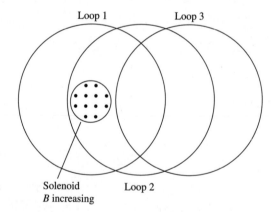

21. A conducting loop around a magnetic field contains two lightbulbs, A and B. The wires connecting the bulbs are ideal, with no resistance. The magnetic field is increasing rapidly.

    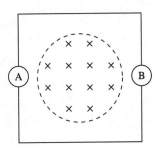

    a. Do the bulbs glow? Why or why not?

    b. If they glow, which bulb is brighter? Or are they equally bright? Explain.

22. A conducting loop around a magnetic field contains three lightbulbs, A, B, and C. The wires connecting the bulbs are ideal, with no resistance. The magnetic field is increasing rapidly. Rank in order, from brightest to least bright, the brightness of the three bulbs.

    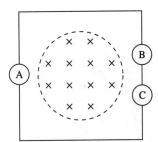

    Order:

    Explanation:

23. A metal wire is resting on a U-shaped conducting rail. The rail is fixed in position, but the wire is free to move.

    a. If the magnetic field is increasing in strength, does the wire:

    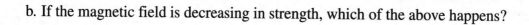

| | |
|---|---|
| i.   Remain in place? | vi.   Move out of the plane of the page, |
| ii.  Move to the right? | breaking contact with the rail? |
| iii. Move to the left? | vii.  Rotate clockwise? |
| iv.  Move up on the page? | viii. Rotate counterclockwise? |
| v.   Move down on the page? | ix.   Some combination of these? If so, which? |

    Explain your choice.

    b. If the magnetic field is decreasing in strength, which of the above happens?

# 33.6 Induced Field and Electromagnetic Waves

# 33.7 Induced Current: Three Applications

No exercises.

## 33.8 Inductors

24. The figure shows the current through an inductor. A positive current is defined as a current going from top to bottom. At the time corresponding to each of the labeled points, does the potential across the inductor (going from top to bottom) increase, decrease, or stay the same?

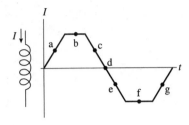

a. _____     e. _____

b. _____     f. _____

c. _____     g. _____

d. _____

25. a. Can you tell which of these inductors has the larger current flowing through it? If so, which one? If not, why not?

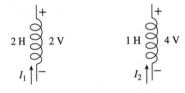

b. Can you tell through which inductor the current is changing most rapidly? If so, which one? If not, why not?

c. If the current enters the inductor from the bottom, can you tell if the current is increasing, decreasing, or staying the same? If so, which one and what is your reasoning? If not, why not?

26. Rank in order, from most positive to most negative, the inductor's potential difference $(\Delta V_L)_a$, $(\Delta V_L)_b$, ..., $(\Delta V_L)_f$, at the six labeled points. $\Delta V_L$ is the change in going from the top of the inductor to the bottom. Some may be equal. Note that $0\ V > -2\ V$.

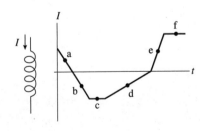

Order:

Explanation:

## 33.9 *LC* Circuits

27. An *LC* circuit oscillates at a frequency of 2000 Hz. What will the frequency be if the inductance is quadrupled?

28. The capacitor in an *LC* circuit has maximum charge at $t = 1$ $\mu$s. The current through the inductor next reaches a maximum at $t = 3$ $\mu$s.

    a. When will the inductor current reach a maximum in the opposite direction?

    b. What is the circuit's period of oscillation?

29. Three *LC* circuits are made with the same capacitor but different inductors. The figure shows the inductor current as a function of time. Rank in order, from largest to smallest, the three inductances $L_1$, $L_2$, and $L_3$. Some may be equal.

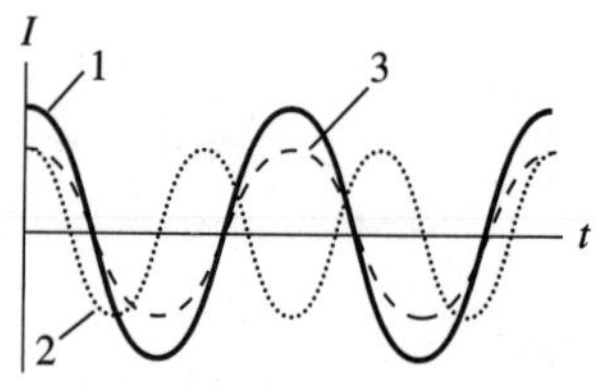

    Order:

    Explanation:

## 33.10 *LR* Circuits

30. Rank in order, from largest to smallest, the three time constants $\tau_1$, $\tau_2$, and $\tau_3$ for these three circuits.

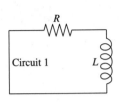 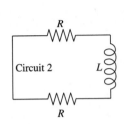 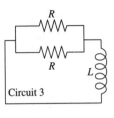

Order:

Explanation:

31. Three *LR* circuits are made with the same resistor but different inductors. The figure shows the inductor current as a function of time. Rank in order, from largest to smallest, the three inductances $L_1$, $L_2$, and $L_3$.

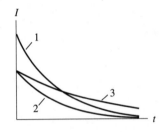

Order:

Explanation:

32. a. What is the battery current immediately after the switch closes? Explain.

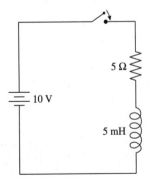

b. What is the battery current after the switch has been closed a long time? Explain.

# 34 Electromagnetic Fields and Waves

## 34.1 Electromagnetic Fields and Forces

1. Draw the electric force vector $\vec{F}_E$, the magnetic force vector $\vec{F}_B$, and the net force vector on the moving charge.

a.

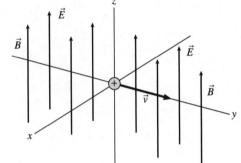

b.

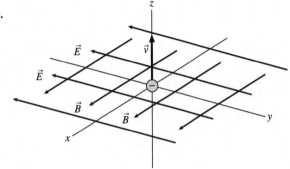

2. Force $\vec{F}$ is the net force on a positive charge moving through a region of space where the electric and magnetic fields are perpendicular to each other.

   a. Draw the magnetic field vectors.

   b. Draw the electric field vectors.

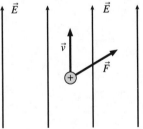

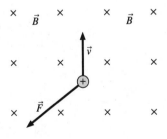

## 34.2  *E* or *B*? It Depends on Your Perspective

3. In frame S, a positive charge moves through the magnetic field shown.
   a. Draw a vector on the charge to show the magnetic force in S.
   b. What are the speed *V* and direction of a reference frame S′ in which there is no magnetic force? Explain.

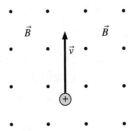

   c. What are the type and direction of any fields in S′ that could cause the observed force on the charge?

4. Sharon drives her rocket through a magnetic field, traveling to the right at a speed of 1000 m/s as measured by Bill. As she passes Bill, she shoots a positive charge backward at a speed of 1000 m/s relative to her.
   a. According to Sharon, what kind of force or forces act on the charge? In which directions? Draw the forces on the charge.

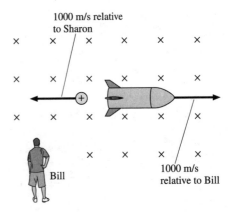

   b. According to Bill, what kind of force or forces act on the charge? In which directions? Explain.

5. In frame S, a positive charge moves to the right at speed $v$. Frame S′ travels to the right at speed $V = v$ relative to S. Frame S″ travels to the right at speed $V = 2v$ relative to S. The figure below shows the charge three times, once in each reference frame.

   a. For each:
      - Draw and label a velocity vector on the charge showing its motion in that frame.
      - Draw and label the electric and magnetic field vectors at the points above and below the charge. Use the notation of circled × and • to show fields into or out of the page.

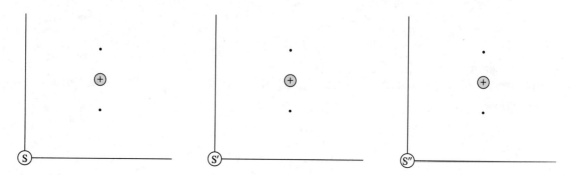

   b. Does it make sense to talk about "the" magnetic field? Why or why not?

## 34.3  Faraday's Law Revisited

## 34.4  The Displacement Current

6. If you curl the fingers of your right hand as shown, is the magnetic flux positive or negative?

a.

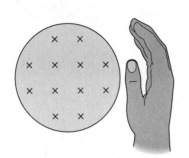

b.

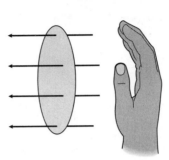

Sign of $\Phi_m$ _____              Sign of $\Phi_m$ _____

7. If you curl the fingers of your right hand as shown, is the emf positive or negative?

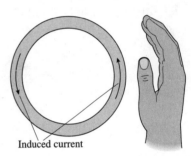

Induced current

8. What is the current through surface S?

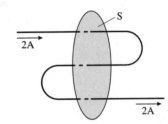

9. The capacitor in this circuit was initially charged, then the switch was closed. At this instant of time, the potential difference across the resistor is $\Delta V_R = 4$ V.

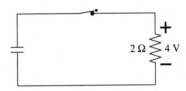

a. At this instant of time, what is the current through the resistor?

b. At this instant of time, what is the current through the space between the capacitor plates?

c. At this instant of time, what is the displacement current through the space between the capacitor plates?

d. Is the displacement current really a current? If so, what are the moving charges? If not, what is the displacement current?

10. Consider these two situations:

a.

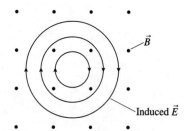

b.

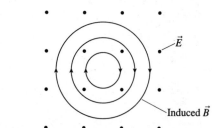

Is the magnetic field strength increasing, decreasing, or not changing? Explain.

Is the electric field strength increasing, decreasing, or not changing? Explain.

11. Consider these two situations:
   a. Draw the induced electric field.

   b. Draw the induced magnetic field.

$\times$ $\quad$ $\times$ $\quad$ $\times$ $\quad$ $\times$

$\times$ $\quad$ $\times$ $\quad$ $\times$ $\quad$ $\times$

$\times$ $\quad$ $\times$ $\quad$ $\times$ $\quad$ $\times$

$\times$ $\quad$ $\times$ $\quad$ $\times$ $\quad$ $\times$

$\vec{B}$-field rapidly increasing

$\times$ $\quad$ $\times$ $\quad$ $\times$ $\quad$ $\times$

$\times$ $\quad$ $\times$ $\quad$ $\times$ $\quad$ $\times$

$\times$ $\quad$ $\times$ $\quad$ $\times$ $\quad$ $\times$

$\times$ $\quad$ $\times$ $\quad$ $\times$ $\quad$ $\times$

$\vec{E}$-field rapidly increasing

## 34.5 Maxwell's Equations

## 34.6 Electromagnetic Waves

## 34.7 Properties of Electromagnetic Waves

12. This is an electromagnetic plane wave traveling into the page. Draw the magnetic field vectors $\vec{B}$ at the dots.

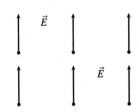

13. This is an electromagnetic wave.
    a. Draw the velocity vector $\vec{v}_{em}$.

    b. Draw $\vec{E}$, $\vec{B}$, and $\vec{v}_{em}$ a half cycle later.

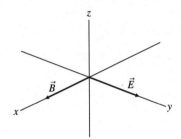

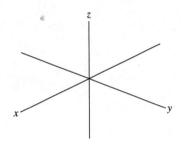

14. Do the following represent possible electromagnetic waves? If not, why not?

a.

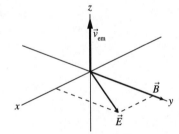

b.

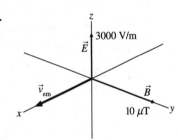

c.

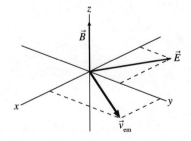

d.

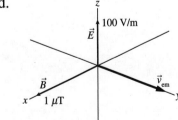

15. The intensity of an electromagnetic wave is 10 W/m$^2$. What will be the intensity if:

a. The amplitude of the electric field is doubled?

b. The amplitude of the magnetic field is doubled?

c. The amplitudes of both the electric field and the magnetic field are doubled?

d. The frequency is doubled?

# 34.8 Polarization

16. A polarized electromagnetic wave passes through a polarizing filter. Draw the electric field of the wave after it has passed through the filter.

a.

b.

17. A polarized electromagnetic wave passes through a series of polarizing filters. Draw the electric field of the wave after it has passed through each filter.

a.

b.

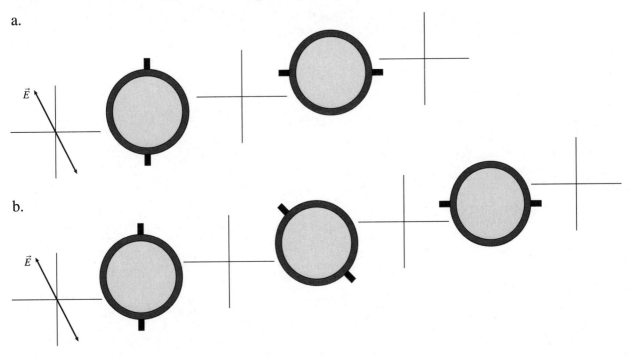

18. The intensity of a polarized electromagnetic wave is 10 W/m². What will be the intensity of the wave after it passes through a polarizing filter whose axis makes the following angle with the plane of polarization?

$\theta = 0°$ _____          $\theta = 60°$ _____

$\theta = 30°$ _____          $\theta = 90°$ _____

$\theta = 45°$ _____

# 35 AC Circuits

## 35.1 AC Sources and Phasors

1. The figure shows emf phasors A, B, and C.
   a. What is the instantaneous value of the emf?

   A _____    B _____    C _____

   b. At this instant, is the magnitude of the emf increasing, decreasing, or holding constant?

   A _____    B _____    C _____

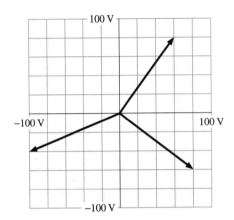

2. Draw a phasor diagram for the following emfs.
   a. $(100 \text{ V})\cos\omega t$ at $\omega t = 240°$    b. $(400 \text{ V})\cos\omega t$ at $t = \frac{1}{3}T$    c. $(200 \text{ V})\cos\omega t$ at $t = 0$

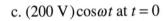

3. The current phasor is shown for a 10 Ω resistor.
   a. What is the instantaneous resistor voltage $v_R$?

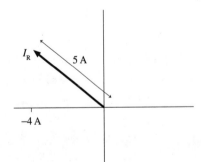

   b. What is the peak resistor voltage $V_R$?

4. The peak current through a resistor is 2.0 A. What is the peak current if:

   a. The resistance $R$ is doubled?

   b. The peak emf $\mathcal{E}_0$ is doubled?

   c. The frequency $\omega$ is doubled?

## 35.2 Capacitor Circuits

5. The peak current through a capacitor is 2.0 A. What is the peak current if:

   a. The peak emf $\mathcal{E}_0$ is doubled?

   b. The capacitance $C$ is doubled?

   c. The frequency $\omega$ is doubled?

6. Current and voltage graphs are shown for a capacitor circuit with $\omega = 1000$ rad/s.

   a. What is the capacitive reactance $X_C$?

   b. What is the capacitance $C$?

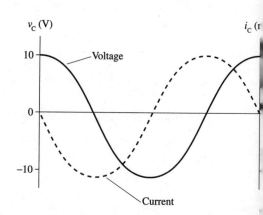

7. A 13 $\mu$F capacitor is connected to a 5.5 V/250 Hz oscillator. What is the instantaneous capacitor current $i_C$ when $\mathcal{E} = -5.5$ V?

8. Consider these three circuits.

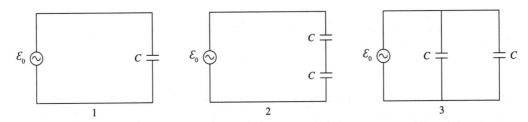

Rank in order, from largest to smallest, the peak currents $(I_C)_1$ to $(I_C)_3$.

Order:

Explanation:

9. Consider these four circuits.

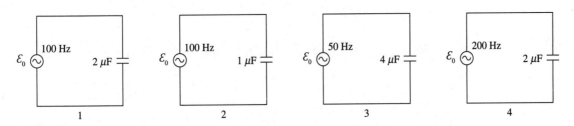

Rank in order, from largest to smallest, the capacitive reactances $(X_C)_1$ to $(X_C)_4$.

Order:

Explanation:

## 35.3 *RC* Filter Circuits

10. A low-pass *RC* filter has a crossover frequency $f_c = 200$ Hz. What is $f_c$ if:
    a. The resistance *R* is doubled?



    b. The capacitance *C* is doubled?



    c. The peak emf $\mathcal{E}_0$ is doubled?



11. What new resistor value *R* will give this circuit the same value of $\omega_c$ if the capacitor value is changed to:

    a. $C = 1\ \mu F$          $R =$ _____

    b. $C = 4\ \mu F$          $R =$ _____

    c. $C = 20\ \mu F$        $R =$ _____

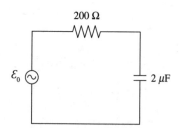

12. Consider these three circuits.

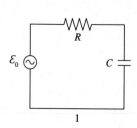 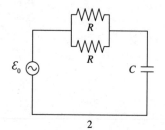 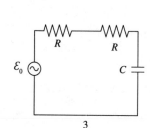

    Rank in order, from largest to smallest, the crossover frequencies $\omega_{c1}$ to $\omega_{c3}$.

    Order:

    Explanation:



13. The text claims that $V_R = V_C = \mathcal{E}_0/\sqrt{2}$ at $\omega = \omega_c$. If this is true, then $V_R + V_C > \mathcal{E}_0$. Is it possible for their sum to be larger than $\mathcal{E}_0$? Explain.

## 35.4 Inductor Circuits

14. The peak current passing through an inductor is 2.0 A. What is the peak current if:
    a. The peak emf $\mathcal{E}_0$ is doubled?

    b. The inductance $L$ is doubled?

    c. The frequency $\omega$ is doubled?

15. Current and voltage graphs are shown for an inductor circuit with $\omega = 1000$ rad/s.
    a. What is the inductive reactance $X_L$?

    b. What is the inductance $L$?

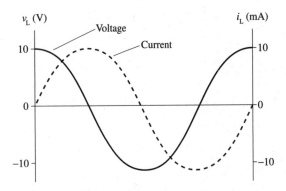

16. Consider these four circuits.

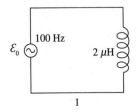 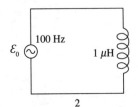 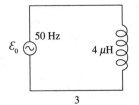 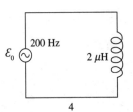

Rank in order, from largest to smallest, the inductive reactances $(X_L)_1$ to $(X_L)_4$.

Order:

Explanation:

## 35.5 The Series *RLC* Circuit

17. The resonance frequency of a series *RLC* circuit is 1000 Hz. What is the resonance frequency if:
    a. The resistance *R* is doubled?

    b. The inductance *L* is doubled?

    c. The capacitance *C* is doubled?

    d. The peak emf $\mathcal{E}_0$ is doubled?

    e. The frequency $\omega$ is doubled?

18. For these combinations of resistance and reactance, is a series *RLC* circuit in resonance (Yes or No)? Does the current lead the emf, lag the emf, or is it in phase with the emf?

| *R* | $X_L$ | $X_C$ | Resonance? | Current? |
|-----|-------|-------|------------|----------|
| 100 Ω | 100 Ω | 50 Ω | _____ | _____ |
| 100 Ω | 50 Ω | 100 Ω | _____ | _____ |
| 100 Ω | 75 Ω | 75 Ω | _____ | _____ |

19. In this series *RLC* circuit, is the emf frequency less than, equal to, or greater than the resonance frequency $\omega_0$? Explain.

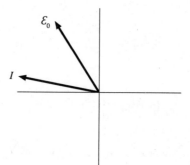

20. The resonance frequency of a series *RLC* circuit is less than the emf frequency. Does the current lead or lag the emf? Explain.

21. Consider these four circuits. They all have the same resonance frequency $\omega_0$.

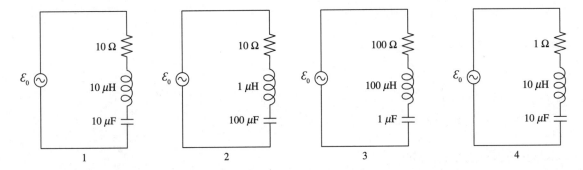

Rank in order, from largest to smallest, the maximum currents $(I_{max})_1$ to $(I_{max})_4$.

Order:

Explanation:

22. The current in a series *RLC* circuit lags the emf by 20°. You cannot change the emf. What two different things could you do to the circuit that would increase the power delivered to the circuit by the emf?

## 35.6 Power in AC Circuits

23. An average power dissipated by a resistor is 4.0 W. What is $P_{avg}$ if:

    a. The resistance $R$ is doubled?

    b. The peak emf $\mathcal{E}_0$ is doubled?

    c. Both are doubled simultaneously?

24. Consider these three circuits.

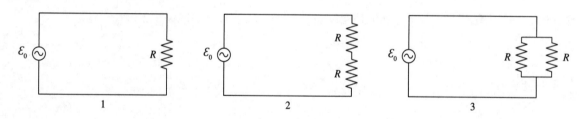

    Rank in order, from largest to smallest, the average powers $P_1$ to $P_3$ delivered by the three emfs.

    Order:

    Explanation:

# 36 Relativity

## 36.1 Relativity: What's It All About?

## 36.2 Galilean Relativity

1. In which reference frame, S or S′, does the ball move faster?

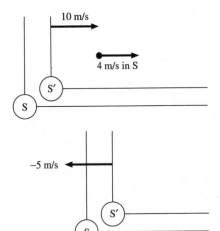

2. Frame S′ moves relative to frame S as shown.

   a. A ball is at rest in frame S′. What are the speed and direction of the ball in frame S?

   b. A ball is at rest in frame S. What are the speed and direction of the ball in frame S′?

3. Frame S′ moves parallel to the x-axis of frame S.

   a. Is there a value of v for which the ball is at rest in S′? If so, what is v? If not, why not?

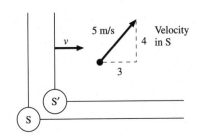

   b. Is there a value of v for which the ball has a minimum speed in S′? If so, what is v? If not, why not?

4. Anjay can swim at a steady speed of 2 mph. He needs to cross a river that flows west to east at 4 mph. Anjay jumps in at point A and swims due north (i.e., his head always points due north) until reaching the opposite shore. Where does Anjay land?

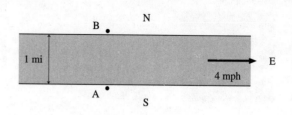

5. a. What are the speed and direction of each ball in a reference frame that moves with ball 1?

   b. What are the speed and direction of each ball in a reference frame that moves with ball 2?

6. What are the speed and direction of each ball in a reference frame that moves to the right at 2 m/s?

# 36.3 Einstein's Principle of Relativity

7. A lighthouse beacon alerts ships to the danger of a rocky coastline.

   a. According to the lighthouse keeper, with what speed does the light leave the lighthouse?

   b. A boat is approaching the coastline at speed 0.5$c$. According to the captain, with what speed is the light from the beacon approaching her boat?

8. As a rocket passes the earth at 0.75$c$, it fires a laser perpendicular to its direction of travel.

   a. What is the speed of the laser beam relative to the rocket?

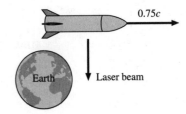

   b. What is the speed of the laser beam relative to the earth?

9. Teenagers Sam and Tom are playing chicken in their rockets. As seen from the earth, each is traveling at 0.95$c$ as he approaches the other. Sam fires a laser beam toward Tom.

   a. What is the speed of the laser beam relative to Sam?

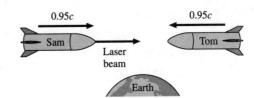

   b. What is the speed of the laser beam relative to Tom?

## 36.4 Events and Measurements

10. It is a bitter cold day at the South Pole, so cold that the speed of sound is only 300 m/s. The speed of light, as always, is 300 m/$\mu$s. A firecracker explodes 600 m away from you.

    a. How long after the explosion until you see the flash of light?_____

    b. How long after the explosion until you see hear the sound?    _____

    c. Suppose you see the flash at $t$ = 2.000002 s. At what time was the explosion? _____

    d. What are the spacetime coordinates for the event "firecracker explodes"? Assume that you are at the origin and that the explosion takes place at a position on the positive $x$-axis.

11. Firecracker 1 is 300 m from you. Firecracker 2 is 600 m from you in the same direction. You see both explode at the same time. Define event 1 to be "firecracker 1 explodes" and event 2 to be "firecracker 2 explodes." Does event 1 occur before, after, or at the same time as event 2? Explain.

12. Firecrackers 1 and 2 are 600 m apart. You are standing exactly halfway between them. Your lab partner is 300 m on the other side of firecracker 1. You see two flashes of light, from the two explosions, at exactly the same instant of time. Define event 1 to be "firecracker 1 explodes" and event 2 to be "firecracker 2 explodes." According to your lab partner, based on measurements he or she makes, does event 1 occur before, after, or at the same time as event 2? Explain.

13. Your clocks and calendars are synchronized with the clocks and calendars in a star system exactly 10 light years from earth that is at rest relative to the earth. You receive a TV transmission from the star system that shows a date and time display. The date it shows is June 17, 2050.

    When you glance over at your own wall calendar, the date it shows is _____.

14. Two trees are 600 m apart. You are standing exactly halfway between them and your lab partner is at the base of tree 1. Lightning strikes both trees.

   a. Your lab partner, based on measurements he or she makes, determines that the two lightning strikes were simultaneous. What did you see? Did you see the lightning hit tree 1 first, hit tree 2 first, or hit them both at the same instant of time? Explain.

   b. Lightning strikes again. This time your lab partner sees both flashes of light at the same instant of time.  What did you see? Did you see the lightning hit tree 1 first, hit tree 2 first, or hit them both at the same instant of time? Explain.

   c. In the scenario of part b, were the lightning strikes simultaneous? Explain.

15. You are at the origin of a coordinate system containing clocks, but you're not sure if the clocks have been synchronized. The clocks have reflective faces, allowing you to read them by shining light on them. You flash a bright light at the origin at the instant your clock reads $t = 2.000000$ s.

   a. At what time will you see the reflection of the light from a clock at $x = 3000$ m?

   b. When you see the clock at $x = 3000$ m, it reads 2.000020 s. Is the clock synchronized with your clock at the origin? Explain.

## 36.5 The Relativity of Simultaneity

16. Two supernovas, labeled L and R, occur on opposite sides of a galaxy, at equal distances from the center. The supernovas are seen at the same instant on a planet at rest in the center of the galaxy. A spaceship is entering the galaxy from the left at a speed of $0.999c$ relative to the galaxy.

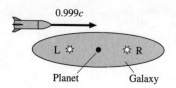

a. According to astronomers on the planet, were the two explosions simultaneous? Explain why.

b. Which supernova, L or R, does the spaceship crew *see* first? _____

c. Did the supernova that was *seen* first necessarily *happen* first in the rocket's frame? Explain.

d. Is "two light flashes reach the planet at the same instant" an event? To help you decide, could you arrange for something to happen only if two light flashes from opposite directions arrive at the same time? Explain.

If you answered Yes to part d, then the crew on the spaceship will also determine, from their measurements, that the light flashes reach the planet at the same instant. (Experimenters in different reference frames may disagree about when and where an event occurs, but they all agree that it *does* occur.)

e. The figure below shows the supernovas in the spaceship's reference frame with the *assumption* that the supernovas are simultaneous. The second half of the figure is a short time after the explosions. Draw two circular wave fronts to show the light from each supernova. Neither wave front has yet reached the planet. Be sure to consider:

   • The points on which the wave fronts are centered.
   • The wave speeds of each wave front.

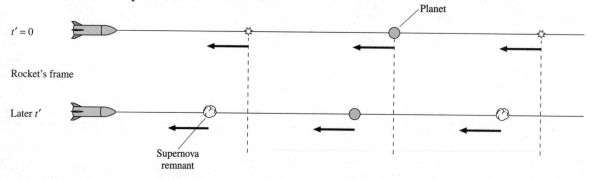

f. According to your diagram, are the two wave fronts going to reach the planet at the same instant of time? Why or why not?

g. Does your answer to part f conflict with your answer to part d? _____

If so, what different assumption could you make about the supernovas in the rocket's frame that would bring your wave-front diagram into agreement with your answer to part d?

h. So according to the spaceship crew, are the two supernovas simultaneous? If not, which happens first?

17. Peggy is standing at the center of her railroad car as it passes Ryan on the ground. Firecrackers attached to the ends of the car explode. A short time later, the flashes from the two explosions arrive at Peggy at the same time.

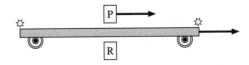

a. Were the explosions simultaneous in Peggy's reference frame? If not, which exploded first? Explain.

b. Were the explosions simultaneous in Ryan's reference frame? If not, which exploded first? Explain.

18. A rocket is traveling from left to right. At the instant it is halfway between two trees, lightning simultaneously (in the rocket's frame) hits both trees.

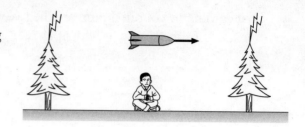

 a. Do the light flashes reach the rocket pilot simultaneously? If not, which reaches him first? Explain.

 b. A student was sitting on the ground halfway between the trees as the rocket passed overhead. According to the student, were the lightning strikes simultaneous? If not, which tree was hit first? Explain.

# 36.6 Time Dilation

19. Clocks $C_1$ and $C_2$ in frame S are synchronized. Clock C′ moves at speed $v$ relative to frame S. Clocks C′ and $C_1$ read exactly the same as C′ goes past. As C′ passes $C_2$, is the time shown on C′ earlier than, later than, or the same as the time shown on $C_2$? Explain.

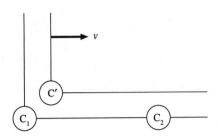

20. Your friend flies from Los Angeles to New York. She carries an accurate stopwatch with her to measure the flight time. You and your assistants on the ground also measure the flight time.

    a. Identify the two events associated with this measurement.

    b. Who, if anyone, measures the proper time?   _____

    c. Who, if anyone, measures the shorter flight time?   _____

    d. Who, if anyone, measures the longer flight time?   _____

21. You're passing a car on the highway. You want to know how much time is required to completely pass the car, from no overlap between the cars to no overlap between the cars. Call your car A and the car you are passing B.

    a. Specify two events that can be given spacetime coordinates. In describing the events, refer to cars A and B and to their front bumpers and rear bumpers.

    b. In either reference frame, is there *one* clock that is present at both events?   _____

    c. Who, if anyone, measures the proper time between the events?   _____

## 36.7 Length Contraction

22. Your friend flies from Los Angeles to New York. He determines the distance using the tried-and-true $d = vt$. You and your assistants on the ground also measure the distance, using meter sticks and surveying equipment.

    a. Who, if anyone, measures the proper length? _____

    b. Who, if anyone, measures the shorter distance? _____

23. Experimenters in B's reference frame measure $L_A = L_B$. Do experimenters in A's reference frame agree that A and B are the same length? If not, which do they find to be longer? Explain.

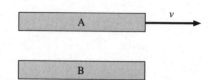

24. As a meter stick flies past you, you simultaneously measure the positions of both ends and determine that $L < 1$ m.

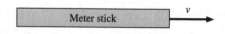

    a. To an experimenter in frame S′, the meter stick's frame, did you make your two measurements simultaneously? If not, which end did you measure first? Explain.

    Hint: Review the reasoning about simultaneity that you used in Exercises 16–18.

    b. Can experimenters in frame S′ give an explanation for why your measurement is < 1 m?

25. A 100-m-long train is heading for an 80-m-long tunnel. If the train moves sufficiently fast, is it possible, according to experimenters on the ground, for the entire train to be inside the tunnel at one instant of time? Explain.

# 36.8 The Lorentz Transformations

26. A rocket travels at speed $0.5c$ relative to the earth.

    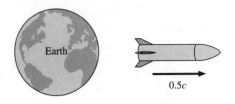

    a. The rocket shoots a bullet in the forward direction at speed $0.5c$ relative to the rocket. Is the bullet's speed relative to the earth less than, greater than, or equal to $c$?

    b. The rocket shoots a second bullet in the backward direction at speed $0.5c$ relative to the rocket. In the earth's frame, is the bullet moving right, moving left, or at rest?

27. The rocket speeds are shown relative to the earth. Is the speed of A relative to B greater than, less than, or equal to $0.8c$?

    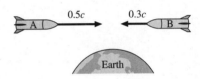

# 36.9 Relativistic Momentum

28. Particle A has half the mass and twice the speed of particle B. Is $p_A$ less than, greater than, or equal to $p_B$? Explain.

29. Particle A has one-third the mass of particle B. The two particles have equal momenta. Is $u_A$ less than, greater than, or equal to $3u_B$? Explain.

30. Event A occurs at spacetime coordinates (300 m, 2 $\mu$s).

    a. Event B occurs at spacetime coordinates (1200 m, 6 $\mu$s). Could A possibly be the cause of B? Explain.

    b. Event C occurs at spacetime coordinates (2400 m, 8 $\mu$s). Could A possibly be the cause of C? Explain.

31. Event B occurs at $t_B = 10.0$ $\mu$s. An earlier event A, at $t_A = 5.0$ $\mu$s, is the cause of B. What is the maximum possible distance that A can be from B?

## 36.10 Relativistic Energy

32. Can a particle of mass $m$ have total energy less than $mc^2$? Explain.

33. Consider these 4 particles:

| Particle | Rest energy | Total energy |
| --- | --- | --- |
| 1 | $A$ | $A$ |
| 2 | $B$ | $2B$ |
| 3 | $2C$ | $4C$ |
| 4 | $3D$ | $5D$ |

Rank in order, from largest to smallest, the particles' speeds $u_1$ to $u_4$.

Order:

Explanation:

# 37 The End of Classical Physics

**Note**: Most of the exercises in this chapter are to give you practice reasoning using *macroscopic* evidence and observations to reason about *microscopic* processes.

## 37.1 Physics in the 1800s

## 37.2 Faraday

## 37.3 Cathode Rays

## 37.4 J. J. Thomson and the Discovery of the Electron

1. Water containing dissolved electrolytes (e.g., salt) is a conductor. *How* does an electric current flow through a liquid? That is, what are the charge carriers? What causes them to move as a current? In which direction do they move?

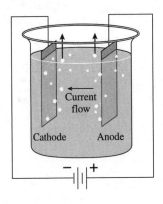

2. Ordinarily, a gas is an insulator. But a potential difference of several hundred volts causes a gas to become a conductor if the pressure is reduced to ≈0.001 atm. We call this a gas discharge, and it is associated with the generation of light. *How* does an electric current flow through a gas? That is, what are the charge carriers? In which direction do they move? Why are low pressure and high voltage necessary?

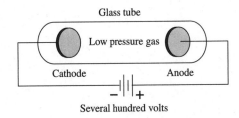

3. Over 100 years ago it was observed that a nearby source of x rays or radioactivity causes a charged electroscope to discharge. Replacing the air around the electroscope by a monatomic gas such as helium does not change the outcome. However, no discharge occurs if the electroscope is in a vacuum.

   a. Do the x rays or radioactivity affect the electroscope itself, or do they affect the environment around the electroscope? Use the observations to explain your reasoning.

   b. What is the *mechanism* by which the discharge occurs?

   c. What can you infer from these observations about the properties of atoms? Possibilities you might want to consider are the size of atoms, the mass of atoms, the divisibility of atoms, the structure of atoms, and perhaps other properties.

4. a. Summarize the experimental evidence *prior* to the research of Thomson by which you might conclude that cathode rays are some kind of particle.

   b. Summarize the experimental evidence *prior* to the research of Thomson by which you might conclude that cathode rays are some kind of wave.

5. Cathode rays cause a green spot to appear on the glass at the front of a Crookes tube. A magnetic field deflects the spot sideways, but the spot remains the *same size*. It does not smear or spread out. This is an interesting observation.

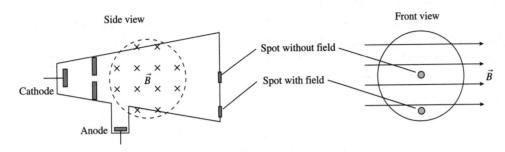

a. Assuming that cathode rays are negatively charged particles, what can you conclude about the charge-to-mass ratios of different cathode-ray particles?

b. What can you conclude about the speeds of different cathode-ray particles?

6. Thomson observed deflection of the cathode-ray particles due to magnetic and electric fields, but there was no observed deflection due to gravity. Why not?

7. In the absence of fields, the green dot caused by the cathode rays appears at the center of the front face of the tube, as shown in the figure of Exercise 5. Prior to Thomson, it was discovered that an electrode in the center of the front face of the tube collects a current of negative charge. Can you conclude from these observations that cathode rays are negatively charged particles? Or is the evidence insufficient to draw this conclusion? Explain your reasoning.

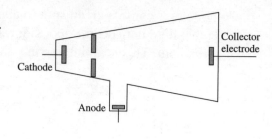

8. Why did Thomson carry out the experiment in which he placed the collecting electrode off to the side of the tube? What was the significance of this experiment?

9. A magnetic field exists in the space between two parallel electrodes.
   a. Sketch the trajectory of the electron if the electrodes are uncharged.
   b. Use plus and minus symbols to show how the electrodes should be charged to achieve zero deflection of the electron.

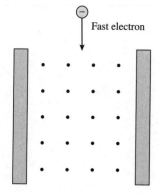

10. Soon after x rays were discovered, it was found that x rays ionize air molecules. Consider an experiment in which x rays are directed into a parallel-plate capacitor.

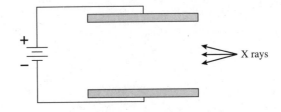

a. On the figure, draw and label the electric field $\vec{E}_{cap}$ of the capacitor.

b. The x-ray ionization creates electrons and positive ions in the space between the electrodes. Use plusses and minuses to show where these charges are most likely to be found.

c. Do the ions and electrons create an electric field? If so, in which direction does it point?

d. What effect does the x-ray ionization of the air have on the *net* electric field between the electrodes?

e. The earliest efforts to deflect cathode rays with an electric field failed. Thomson succeeded, but only by reducing the pressure in the cathode-ray tube to the lowest possible levels. Explain these observations.

11. Thomson found that $(q/m)_{\text{cathode rays}} \approx 1000(q/m)_{\text{H-ion}}$. (The charge-to-mass ratio of the hydrogen ion $H^+$ was known from electrolysis.) This suggests either:

i.   $q_{\text{cathode rays}} \approx 1000 q_{\text{H-ion}}$,

ii.  $m_{\text{cathode rays}} \approx \frac{1}{1000} m_{\text{H-ion}}$, or

iii. Some combination of these.

In 1897, when Thomson made his measurements, whole atoms were thought to be the smallest unit of matter and Millikan had not yet established that there is a fundamental unit of charge. So why did Thomson reject option i? Cite specific experimental evidence.

12. Thomson used cathodes of aluminum, iron, platinum, and other metals. He found the same value of $(q/m)_{\text{cathode rays}}$ for each. What conclusion can you draw from this observation?

# 37.5  Millikan and the Fundamental Unit of Charge

13. The figure shows a positively charged oil drop held at rest between two electrodes.

   a. Draw a free-body diagram for the oil drop while it is held at rest.

   b. If the electric field is turned off, the drop begins to fall very slowly at constant speed. Draw a free-body diagram of the drop as it falls.

14. Millikan adjusted the potential difference between the electrodes of his experiment by sliding a wire up and down a high-resistance wire that was connected to a battery.

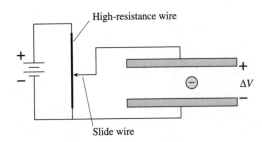

   a. Why does $\Delta V$ change as the point of contact is moved?

   b. A negatively charged drop is falling. To stop its motion and keep it at rest, should the slide wire be moved up or down? Explain.

15. Millikan observed both positive and negative oil drops.

   a. Why didn't all the oil drops have the same sign?

   b. Did the fact that he observed both positive and negative drops affect his conclusion that he had measured the "fundamental unit" of charge? Why or why not?

16. Charge measurements on six oil drops give the following values, all in some unknown units of charge.

| Drop | $q$ |
|------|------|
| 1 | +7.0 |
| 2 | +24.5 |
| 3 | −7.0 |
| 4 | −28.0 |
| 5 | +14.0 |
| 6 | −10.5 |

In these units, what is the fundamental unit of charge? Explain.

## 37.6  Rutherford and the Discovery of the Nucleus

## 37.7  Into the Nucleus

17. A block of uranium emits alpha particles continuously. As Rutherford showed, each alpha particle has a charge and a mass.

    a. Does the continuous emission of these particles violate the law of conservation of mass? If not, why not?

    b. Does the continuous emission of these particles violate the law of conservation of charge? If not, why not?

18. A proton gains 50 eV of kinetic energy by moving through a potential difference. What is $\Delta V$?

19. An electron, a proton, and a helium nucleus (He$^{++}$) accelerate through the 10 V potential difference between two capacitor plates. For each, what is the particle's increase in kinetic energy $\Delta K$ in electron volts?

    Electron        $\Delta K = $ _____

    Proton          $\Delta K = $ _____

    He$^{++}$        $\Delta K = $ _____

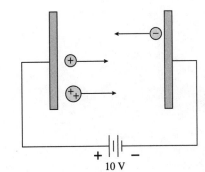

20. Rutherford studied alpha particles using the same crossed-field technique that Thomson had invented to study cathode rays. Assuming that $v_{alpha} \approx v_{cathode\ ray}$ (which turns out to be true), would the deflection of an alpha particle by a magnetic field be larger than, smaller than, or the same as the deflection of a cathode-ray particle by the same field? Explain.

21. Once Thomson showed that atoms consist of very light negative electrons and a much more massive positive charge, why didn't physicists immediately consider a solar-system model of electrons orbiting a positive nucleus? Why would physicists in 1900 object to such a model?

22. Suppose you throw a small, hard rubber ball through a tree. The tree has many outer leaves, so you cannot see clearly into the tree. Most of the time your ball passes through the tree and comes out the other side with little or no deflection. On occasion, the ball emerges at a very large angle to your direction of throw. On rare occasions, it even comes straight back toward you. From these observations, what can you conclude about the structure of the tree? Be specific as to how you arrive at these conclusions *from the observations*.

23. Beryllium is the fourth element in the periodic table. Draw pictures similar to Figure 37.15 showing the structure of neutral Be, of $Be^+$, of $Be^{++}$, and of the negative ion $Be^-$.

Be                    $Be^+$                    $Be^{++}$                    $Be^-$

24. The element hydrogen has three isotopes. The most common has $A = 1$. A rare form of hydrogen (called *deuterium*) has $A = 2$. An unstable, radioactive form of hydrogen (called *tritium*) has $A = 3$. Draw pictures similar to Figure 37.18 showing the structure of these three isotopes. Show all the electrons, protons, and neutrons of each.

$A = 1$                    $A = 2$                    $A = 3$

25. Tritium, the $A = 3$ isotope of hydrogen, is radioactive. One of the neutrons undergoes the transformation $n \rightarrow p^+ + e^- + v$, where $v$ is a massless, chargeless subatomic particle called a *neutrino*. Both the electron and the neutrino are ejected from the nucleus at high speed, but the proton remains. This is called *beta decay* of a nucleus.

a. Draw a picture of the structure of a tritium atom immediately before and immediately after it undergoes beta decay. Show all the electrons, protons, and neutrons.

b. Identify the element, the isotope (the $A$-value), and the charge state (neutral or singly charged) of the atom after the decay. Give your answer in symbolic form, such as $^6Li^+$.

26. Identify the element, the isotope, and the charge state. Give your answer in symbolic form, such as $^4He^+$ or $^8Be^-$.

a.

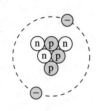

b.

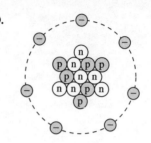

## 37.8 The Emission and Absorption of Light

27. A photograph of an absorption spectrum appears white with black lines. An emission spectrum appears black with bright lines. Why are they different?

28. Light is emitted from a discharge tube only if current flows through the gas. The light immediately ceases if the current is interrupted for any reason. What conclusions can you draw about the *mechanism* by which the light is emitted? (You may want to refer back to Exercise 2.)

# 38 Quantization

## 38.1 The Photoelectric Effect

## 38.2 Einstein's Explanation

1. a. A negatively charged electroscope can be discharged by shining an ultraviolet light on it. How does this happen?

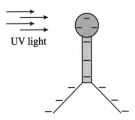

UV light

b. You might think that an ultraviolet light shining on an initially uncharged electroscope would cause the electroscope to become positively charged as photoelectrons are emitted. In fact, ultraviolet light has no noticeable effect on an uncharged electroscope. Why not?

2. In the photoelectric effect experiment, a current is measured while light is shining on the cathode. But this does not appear to be a complete circuit, so how can there be a current? Explain.

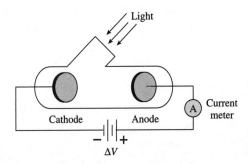

3. Draw the trajectories of several typical photoelectrons when (a) $\Delta V = V_{anode} - V_{cathode} > 0$, (b) $-V_{stop} < \Delta V < 0$, and (c) $\Delta V < -V_{stop}$.

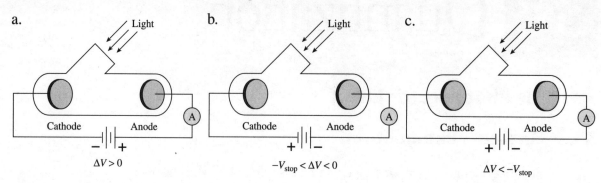

a.
Light
Cathode   Anode
$\Delta V > 0$

b.
Light
Cathode   Anode
$-V_{stop} < \Delta V < 0$

c.
Light
Cathode   Anode
$\Delta V < -V_{stop}$

4. Why are only electrons emitted in the photoelectric effect experiment? Why are positive ions never emitted?

5. The work function of a metal measures:
   i.   The kinetic energy of the electrons in the metal.
   ii.  How tightly the electrons are bound within the metal.
   iii. The amount of work done by the metal when it expands.

   Which of these (perhaps more than one) are correct? Explain.

6. The figure shows the typical photoelectric behavior of a metal as the anode-cathode potential difference $\Delta V$ is varied.

   a. Why do the curves become horizontal for $\Delta V > 0$ V? Shouldn't the current increase as the potential difference increases? Explain.

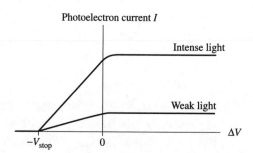

Photoelectron current $I$

Intense light

Weak light

$-V_{stop}$   0   $\Delta V$

   b. Why doesn't the current immediately drop to zero for $\Delta V < 0$ V? Shouldn't $\Delta V < 0$ V prevent the electrons from reaching the anode? Explain.

c. The current is zero for $\Delta V < -V_{stop}$. Where do the photoelectrons go? Are no photoelectrons emitted if $\Delta V < -V_{stop}$? Or if they are, why is there no current? Explain.

7. a. What is the significance of $V_{stop}$? That is, what have you learned if you measure $V_{stop}$?

   **Note:** Don't say that "$-V_{stop}$ is the potential that causes the current to stop." That is merely the definition of $V_{stop}$. It doesn't say what the *significance* of $V_{stop}$ is.

   b. Why is it surprising that $V_{stop}$ is independent of the light intensity? What would you *expect* $V_{stop}$ to do as the intensity increases? Explain.

   c. If the wavelength of the light in a photoelectric effect experiment is increased, does $V_{stop}$ increase, decrease, or stay the same? Explain.

8. Metal 1 has a larger work function than metal 2. Both are illuminated with the same short-wavelength ultraviolet light. Do photoelectrons from metal 1 have a higher speed, a lower speed, or the same speed as photoelectrons from metal 2? Explain.

9. Consider the following characteristics of the photoelectric effect:
   i.   The generation of photoelectrons.
   ii.  The existence of a threshold frequency.
   iii. The photoelectric current increases with increasing light intensity.
   iv.  The photoelectric current is independent of $\Delta V$ for $\Delta V > 0$.
   v.   The photoelectric current decreases slowly as $\Delta V$ becomes more negative.
   vi.  The stopping potential is independent of the light intensity.
   vii. The photoelectron current appears instantly when the light is turned on.

   Which of these *cannot* be explained by classical physics?

10. The figure shows a typical current-versus-potential difference graph for a photoelectric effect experiment. On the figure, draw and label graphs for the following three situations:
    i.   The light intensity is increased.
    ii.  The light frequency is increased.
    iii. The cathode work function is increased.

    In each case, no other parameters of the experiment are changed.

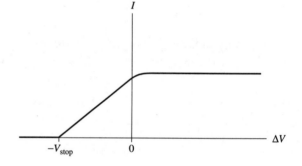

11. The figure shows a typical current-versus-frequency graph for a photoelectric effect experiment. On the figure, draw and label graphs for the following three situations:
    i.   The light intensity is increased.
    ii.  The anode-cathode potential difference is increased.
    iii. The cathode work function is increased.

    In each case, no other parameters of the experiment are changed.

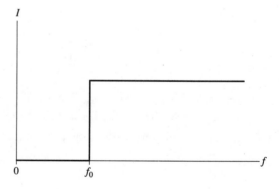

12. The figure shows a typical stopping potential-versus-frequency graph for a photoelectric effect experiment. On the figure, draw and label graphs for the following two situations:

    i.   The light intensity is increased.
    ii.  The cathode work function is increased.

    In each case, no other parameters of the experiment are changed.

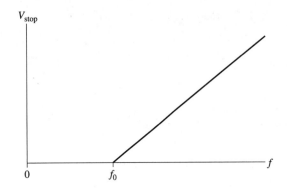

13. In a photoelectric effect experiment, the frequency of the light is increased. As a result:

    i.   There are more photoelectrons.
    ii.  The photoelectrons are faster.

    iii. Both i and ii.
    iv.  Neither i nor ii.

    Explain your choice.

14. In a photoelectric effect experiment, the intensity of the light is increased. As a result:

    i.   There are more photoelectrons.
    ii.  The photoelectrons are faster.

    iii. Both i and ii.
    iv.  Neither i nor ii.

    Explain your choice.

15. A gold cathode (work function = 5.1 eV) is illuminated with light of wavelength 250 nm. It is found that the photoelectron current is zero when $\Delta V = 0$ V. Would the current change if:

    a. The intensity is doubled?

    b. The anode-cathode potential difference is increased to $\Delta V = 5.5$ V?

    c. The cathode is changed to aluminum (work function = 4.3 eV)?

## 38.3 Photons

16. When we say that a photon is a "quantum of light," what does that mean? What is quantized?

17. The intensity of a beam of light is increased but the light's frequency is unchanged. As a result:

    i.   The photons travel faster.
    ii.  Each photon has more energy.
    iii. The photons are larger.
    iv.  There are more photons per second.

    Which of these (perhaps more than one) are true? Explain.

18. The frequency of a beam of light is increased but the light's intensity is unchanged. As a result:

    i.   The photons travel faster.
    ii.  Each photon has more energy.
    iii. There are fewer photons per second.
    iv.  There are more photons per second.

    Which of these (perhaps more than one) are true? Explain.

19. Photons are sometimes represented by pictures like this. Just what is this a picture of? Explain what this "graph" shows.

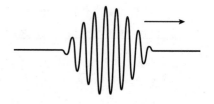

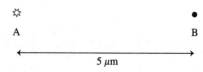

20. Light of wavelength $\lambda = 1\ \mu m$ is emitted from point A. A photon is detected 5 $\mu$m away at point B. On the figure, draw the trajectory that a photon follows between points A and B.

# 38.4  Matter Waves and Energy Quantization

21. Electron 1 is accelerated from rest through a potential difference of 100 V. Electron 2 is accelerated from rest through a potential difference of 200 V. Afterward, which electron has the larger de Broglie wavelength? Explain.

22. An electron and a proton are each accelerated from rest through a potential difference of 100 V. Afterward, which particle has the larger de Broglie wavelength? Explain.

23. Neutron beam 1 has a temperature of 500 K. Neutron beam 2 has a temperature of 5 K. Which neutrons have the larger de Broglie wavelength? Explain.

24. A neutron is shot straight up with an initial speed of 100 m/s. As it rises, does its de Broglie wavelength increase, decrease, or not change? Explain.

25. Double-slit interference of electrons occurs because:
    i.   The electrons passing through the two slits repel each other.
    ii.  Electrons collide with each other behind the slits.
    iii. Electrons collide with the edges of the slits.
    iv.  Each electron goes through both slits.
    v.   The energy of the electrons is quantized.
    vi.  Only certain wavelengths of the electrons fit through the slits.

    Which of these (perhaps more than one) are correct? Explain.

26. Can an electron with a de Broglie wavelength of 2 $\mu$m pass through a slit that is 1 $\mu$m wide?

27. The figure shows the standing de Broglie wave of a particle in a box.

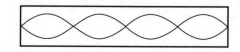

    a. What is the quantum number? _____

    b. Can you determine from this picture whether the "classical" particle is moving to the right or to the left? If so, which is it? If not, why not?

28. A particle in a box of length $L_a$ has $E_1 = 2$ eV. The same particle in a box of length $L_b$ has $E_2 = 50$ eV. What is the ratio $L_a/L_b$?

29. The figure shows a standing de Broglie wave.

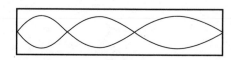

    a. Does this standing wave represent a particle that travels back and forth between the boundaries with a constant speed or a changing speed? Explain.

    b. If the speed is changing, at which end is the particle moving faster and at which end is it moving slower?

# 38.5  Bohr's Model of Atomic Quantization

# 38.6  The Bohr Hydrogen Atom

# 38.7  The Hydrogen Spectrum

30. J. J. Thomson studied the ionization of atoms in collisions with electrons. He accelerated electrons through a potential difference, shot them into a gas of atoms, then used a mass spectrometer to detect any ions produced in the collisions. By using different gases, he found that he could produce singly ionized atoms of all the elements that he tried. When he used higher accelerating voltages, he was able to produce doubly ionized atoms of all elements *except* hydrogen.

   a. Why did Thomson have to use higher accelerating voltages to detect doubly ionized atoms than to detect singly ionized atoms?

   b. What conclusion or conclusions about hydrogen atoms can you draw from these observations? Be specific as to how your conclusions are related to the observations.

31. If an electron is in a *stationary state* of an atom, is the electron at rest? If not, what does the term mean?

32. The diagram shows the energy level diagram of element X.

   a. What is the ionization energy of element X?

$E = 0$ eV -------------

−1 eV ——————— $n = 3$

1240 nm

−2 eV ——————— $n = 2$

−4 eV ——————— $n = 1$

   b. An atom in the ground state absorbs a photon, then emits a photon with a wavelength of 1240 nm. What conclusion can you draw about the energy of the photon that was absorbed?

   c. An atom in the ground state has a collision with an electron, then emits a photon with a wavelength of 1240 nm. What conclusion can you draw about the initial kinetic energy of the electron?

33. The figure shows a hydrogen atom, with an electron orbiting
a proton.

    a. What force or forces act on the electron?

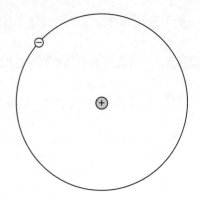

    b. On the figure, draw the electron's velocity, acceleration, and
    the force vectors.

34. Bohr did not include the gravitational force in his analysis of the hydrogen atom. Is this one of
the reasons that his model of the hydrogen atom had only limited success? Explain.

35. a. The stationary state of hydrogen shown on the left has quantum number $n =$ _____

    b. On the right, draw the stationary state of the $n - 1$ state.

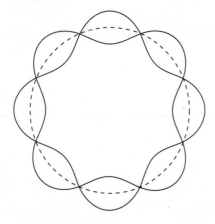 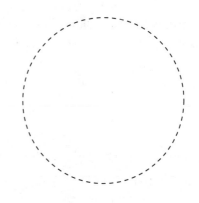

36. The $n = 3$ state of hydrogen has $E_3 = -1.51$ eV.

    a. Why is the energy negative?

    b. What is the physical significance of the specific number 1.51 eV?

37. Why is there no stationary state of hydrogen with $E = -9$ eV?

38. Draw and label an energy level diagram for hydrogen. On it, show all the transitions by which an electron in the $n = 4$ state could emit a photon.

39. a. Which states of a hydrogen atom can be excited by collision with an electron with kinetic energy $K = 12.5$ eV? Explain.

   b. After the collision, the electron:
      i.   Bounces off with $K > 12.5$ eV.
      ii.  Bounces off with $K = 12.5$ eV.
      iii. Bounces off with $K < 12.5$ eV.
      iv.  Has been absorbed.
      Explain your choice.

   c. After the collision, the atom emits a photon. List all the possible $n \to m$ transitions that might occur as a result of this collision.

40. The longest wavelength in the Balmer series is 656 nm.

    a. What transition is this?

    b. If light of this wavelength shines on a container of hydrogen atoms, will the light be absorbed? Why or why not?

41. a. How many electrons, protons, and neutrons are in a hydrogen-like $^{12}$C ion?

    b. Draw a picture of a hydrogen-like $^{12}$C ion, showing all the particles you identified in part a.

# 39 Wave Functions and Uncertainty

## 39.1 Waves, Particles, and the Double-Slit Experiment

## 39.2 Connecting the Wave and Photon Views

1. An experiment has three possible outcomes: A, B, and C. The probability of A is 20% and the probability of B is 30%.

   a. What is the probability that the outcome is either A or B?  _____

   b. What is the probability that the outcome is C?  _____

   c. What is the probability that the outcome is not B?  _____

   d. Out of 1000 trials, how many do you *expect* to have the outcome A? _____

   e. If 1000 trials yielded 210 A's, would you be concerned that the stated probabilities are in error? Why or why not?

2. a. The graph shows the linear mass density of a string whose density is a function of position. Which end of the string (left or right) is more massive?

   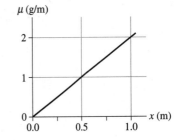

   b. What is the mass of a 1-mm-long segment of string at:

   $x = 0.25$ m    mass = _____

   $x = 0.50$ m    mass = _____

   $x = 0.75$ m    mass = _____

   c. This graph shows the probability density for photons to be detected on the $x$-axis. At which end of the $0$ m $\leq x \leq 1$ m interval ($x \approx 0$ m or $x \approx 1$ m) is a photon more likely to be detected?

   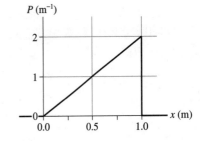

   d. 1,000,000 photons are detected. What is the expected number of photons in a 1-mm-wide interval at:

   $x = 0.25$ m    # photons = _____

   $x = 0.50$ m    # photons = _____

   $x = 0.75$ m    # photons = _____

**39-1**

3. An experiment with light produces 1.0-cm-wide interference fringes. The probability of a photon landing in a 0.01-cm-wide interval at the position $x_1$ is Prob(in 0.01 cm at $x_1$) = 0.030.

   a. What is Prob(in 0.005 cm at $x_1$)?

   b. Can the reasoning you used in part a be extended to find Prob(in 0.5 cm at $x_1$)? If so, what is it? If not, why not?

   c. What is the probability density at $x_1$?

4. Light passes through a single narrow slit. The top graph in the figure shows the measured light intensity $I(x)$ on a screen behind the slit.

   a. Graph the amplitude function $A(x)$. Keep in mind that the amplitude function *oscillates* between positive and negative values.

   b. In the space below your amplitude graph, draw a dot pattern showing the likely distribution of the first ≈100 photons.

   c. What are the most probable value or values of $x$ for detecting a photon?

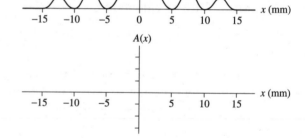

   d. What are the least probable value or values of $x$ for detecting a photon?

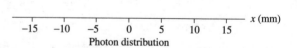

Photon distribution

   e. 1,000,000 photons are detected. Make a rough estimate of the number that are detected in the interval −5 mm < $x$ < +5 mm. Explain your reasoning.

# 39.3  The Wave Function

5. The figure shows the wave function of electrons arriving on a screen along the *x*-axis.

   a. Draw a graph of $|\psi(x)|^2$.

   b. Draw a dot pattern showing the likely distribution of electrons on the screen.

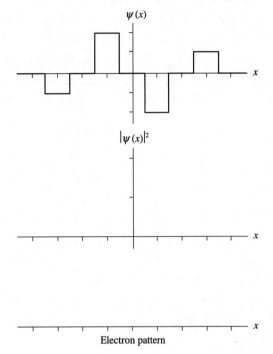

6. The figure shows the wave function of electrons arriving on a screen along the *x*-axis.

   a. Draw a graph of $|\psi(x)|^2$.

   b. Draw a dot pattern showing the likely distribution of electrons on the screen.

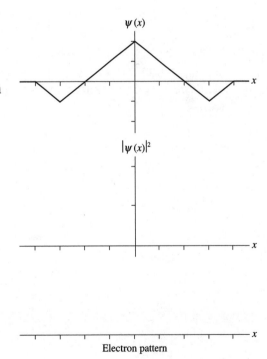

7. The figure shows the dot pattern of electrons landing on a screen.

   a. Draw a graph of the probability density $P(x) = |\psi(x)|^2$ as a function of $x$.

   b. Draw a graph of the electron wave function $\psi(x)$. Keep in mind that wave functions *oscillate* between positive and negative values.

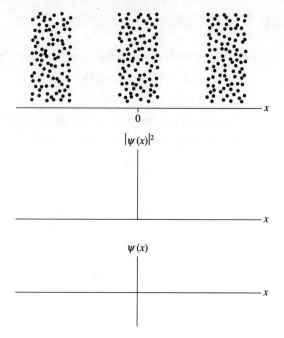

8. The figure shows the dot pattern of electrons landing on a screen.

   a. Draw a graph of the probability density $P(x) = |\psi(x)|^2$ as a function of $x$.

   b. Draw a graph of the electron wave function $\psi(x)$. Keep in mind that wave functions *oscillate* between positive and negative values.

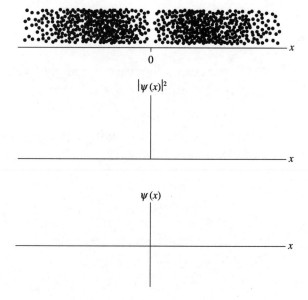

# 39.4 Normalization

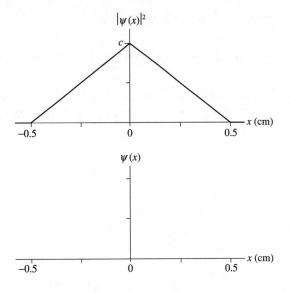

9. The figure shows a graph of $|\psi(x)|^2$.

    a. What value of the constant $c$ makes this a
       normalized wave function? Explain.

    b. What is Prob(in 0.001 cm at $x = 0.25$ cm)?

    c. What is Prob($0$ cm $\leq x \leq 0.5$ cm)?

    d. Draw a graph of the wave function $\psi(x)$. Add an appropriate vertical scale to your graph.
       (Think carefully about the proper shape.)

10. The figure shows the graph of a wave
    function $\psi(x)$.

    a. Draw the graph of $|\psi(x)|^2$. Add an appropriate
       vertical scale to your graph.

    b. Is this a normalized wave function? Explain.

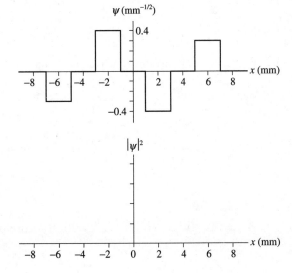

    c. What is the value of:   Prob(5 mm $\leq x \leq 10$ mm)?   _____

                              Prob(−5 mm $\leq x \leq 5$ mm)?   _____

                              Prob(−10 mm $\leq x \leq 10$ mm)? _____

## 39.5 Wave Packets

## 39.6 The Heisenberg Uncertainty Principle

11. Approximately what range of frequencies must be superimposed to create this wave packet?

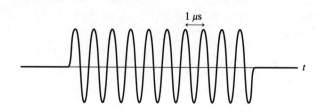

12. A telecommunication system uses pulses with a duration of 5 ns.

   a. About how many such pulses can be transmitted per second?

   b. What is the minimum bandwidth needed to transmit these pulses?

13. Wave packets are shown for particles 1, 2, and 3.

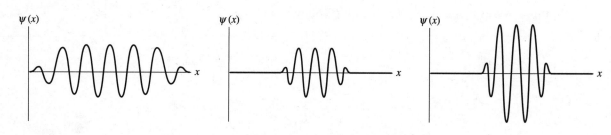

Which particle can have its velocity known most precisely? Explain.

# 40 One-Dimensional Quantum Mechanics

## 40.1  Schrödinger's Equation: The Law of Psi

## 40.2  Solving the Schrödinger Equation

1. The graphs of several functions are shown below. For each, would this function be an acceptable wave function in quantum mechanics? If not, what is wrong with it?

a.
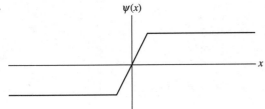

b.
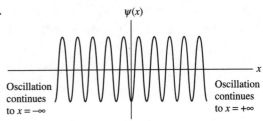

Oscillation continues to $x = -\infty$       Oscillation continues to $x = +\infty$

c.
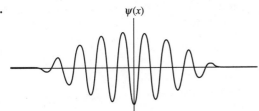

d.
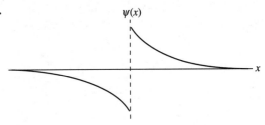

e.
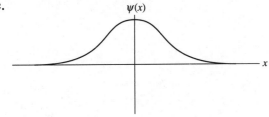

f.
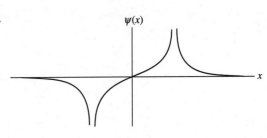

# 40.3  A Particle in a Rigid Box: Energies and Wave Functions

# 40.4  A Particle in a Rigid Box: Interpreting the Solution

# 40.5  The Correspondence Principle

2. Use the properties of waves to explain why a particle in a rigid box can have only certain energies but not others.

3. The graph shows the energy levels and wave functions of the $n = 2$ and $n = 4$ states of a particle in a box.

   a. In which state does the particle have more energy?

   b. In which state are you more likely to find the particle at $x = \frac{3}{4}L$? Explain.

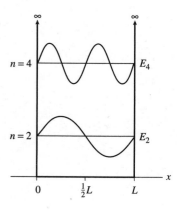

4. Can an atomic particle be at rest in the center of a microscopic box? If so, describe a procedure by which you could arrange this. If not, why not?

5. The correspondence principle says that the *average* behavior of a quantum system should begin to look like the Newtonian solution in the limit that the quantum number becomes very large. What is meant by "the *average* behavior" of a quantum system?

# 40.6  Finite Potential Wells

6. A particle in a potential well is in the $n = 5$ quantum state. How many peaks are there in the probability density $P(x) = |\psi(x)|^2$?

7. What is the quantum number for this particle in a finite potential well? How can you tell?

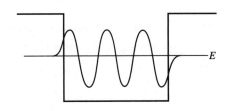

8. The energy-level diagram shows a neutron in a nucleus.
   a. How many possibilities are there for the emission of a gamma-ray photon?

   b. What is the energy of each gamma-ray photon that could be emitted?

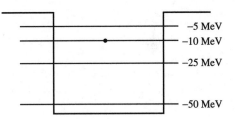

9. Rank in order, from largest to smallest, the penetration distances $\eta_1$ to $\eta_3$ of the wave functions corresponding to these three energy levels.

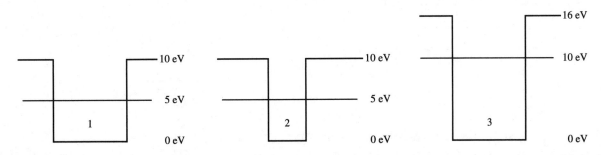

Order:

Explanation:

## 40.7 Wave-Function Shapes

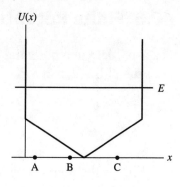

10. Energy $E$ is an allowed energy for a particle in this potential well. Three positions—A, B, and C—are marked on the $x$-axis.

   a. Rank in order, from largest to smallest, the classical kinetic energies of the particle at these three positions.

   b. Rank in order, from largest to smallest, the particle's de Broglie wavelengths at these three positions.

   c. Rank in order, from largest to smallest, the amount of time a classical particle spends traversing an interval of width $\delta x$ at each of these three points.

   d. Rank in order, from largest to smallest, the spacings between the zeros of the wave function in the regions near each of these points. Assume that $n \gg 1$ and that the wave function oscillates many times between the two walls of the potential energy. Explain your ranking.

   e. Rank in order, from largest to smallest, the amplitudes of the wave function in the regions near each of these points. Explain your ranking.

11. The figure shows the $n = 4$ and the $n = 8$ energy levels of a particle in a potential well. Based on your answers to Exercise 10, sketch plausible $n = 4$ and $n = 8$ wave functions.

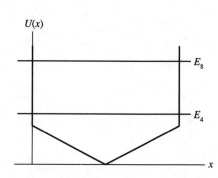

12. The figure shows a solution of the Schrödinger equation.

    a. What is the value of the quantum number $n$? Explain how you know.

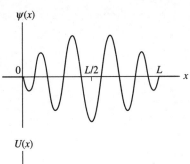

    b. What is the probability of finding the particle at $x < 0$?

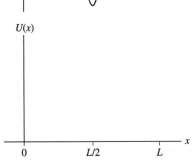

    What is the probability of finding the particle at $x > L$?

    c. At what value or values of $x$ is the particle most likely to be found?

    d. In what general region is the de Broglie wavelength the longest?   _____

    In what general regions is the de Broglie wavelength the shortest?   _____

    e. In what general regions would a classical particle move the fastest?   _____

    In what general region would a classical particle move the slowest?   _____

    f. In what region(s) would a classical particle have maximum kinetic energy?   _____

    In what region(s) would a classical particle have minimum kinetic energy?   _____

    g. Now, on the lower set of axes, sketch a plausible potential well in which the particle is confined.

13. Consider the potential well shown.

    a. Describe how a classical particle with energy $E_a$ would move. In particular, specify where it speeds up, where it slows down, where the turning points are, and which regions are forbidden to the particle.

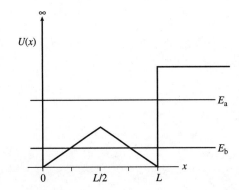

   b. Describe the possible motions of a classical particle with energy $E_b$.

   c. Now consider a quantum particle with energy $E_a$. What is the value of the wave function at $x = 0$?

     Can you determine the value of the wave function at $x = L$? If so, what is it? If not, why not?

     Sketch—on the $E_a$ energy level in the figure—a plausible wave function for energy $E_a$.

   d. Suppose the quantum particle has energy $E_b$. In what region or regions might you find a quantum particle that would be forbidden to a classical particle?

     Sketch—on the $E_b$ energy level—a plausible wave function for energy $E_b$.

## 40.8 The Quantum Harmonic Oscillator

## 40.9 More Quantum Models

14. The figure shows the $n = 2$ energy level and wave function for a quantum harmonic oscillator.

    a. Can you determine, from the figure, the amplitude of the wave function? If so, what is it? If not, why not?

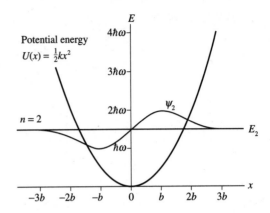

    b. On the figure, identify and label the classical turning points for a particle with energy $E_2$.

    c. On the figure, shade the area between the wave function and the $E_2$ line to show where the wave function penetrates into a classically forbidden region.

15. The figure shows two possible wave functions for an electron in a linear triatomic molecule. Which of these is a bonding orbital and which is an antibonding orbital? Explain how you can distinguish them.

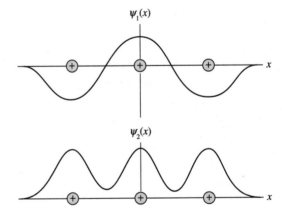

# 40.10 Quantum-Mechanical Tunneling

16. For the potential-energy functions shown below, can a particle on the left side of the potential well ($x < 0.45L$) with energy $E$ tunnel into the right side of the potential well ($x > 0.55L$)? If not, why not?

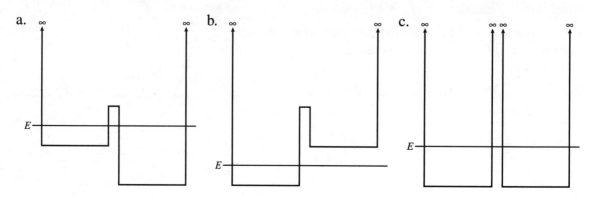

17. Four quantum particles with energy $E$ approach a potential energy barrier from the left. Each has a probability for tunneling through the barrier. Rank in order, from largest to smallest, the tunneling probabilities $(P_{\text{tunnel}})_1$ to $(P_{\text{tunnel}})_4$.

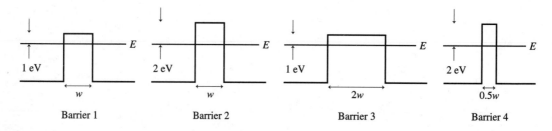

Order:

Explanation:

# 41 Atomic Physics

## 41.1  The Hydrogen Atom: Angular Momentum and Energy

## 41.2  The Hydrogen Atom: Wave Functions and Probabilities

1. List all possible states of a hydrogen atom that have $E = -1.51$ eV.

  | $n$ | $l$ | $m$ |

2. What are the $n$ and $l$ values of the following states of a hydrogen atom?

State = $4d$     $n = $ _____     $l = $ _____

State = $5f$     $n = $ _____     $l = $ _____

State = $6s$     $n = $ _____     $l = $ _____

3. How would you label the hydrogen-atom states with the following quantum numbers?

$(n, l, m) = (4, 3, 0)$     Label = _____

$(n, l, m) = (3, 2, 1)$     Label = _____

$(n, l, m) = (3, 2, -1)$     Label = _____

4. Consider the two hydrogen-atom states $5d$ and $4f$. Which has the higher energy? Explain.

5. a. As a multiple of $\hbar$, what is the angular momentum of a $d$ electron? _____

   b. What is the *maximum* $z$-component of angular momentum of a $d$ electron? _____

   c. Is $(L_z)_{\text{max}}$ greater than, less than, or equal to $L$? _____

   d. What is the significance of your answer to part c?

6. Draw a picture similar to Figure 41.3 showing all possible orientations of the angular momentum vector for a $p$ electron.

7. What is the difference between the *probability density* and the *radial probability density*?

8. Consider a $2s$ electron, as portrayed in Figures 41.5 and 41.6. In your own words, describe how these figures suggest a "shell structure" of electrons around the nucleus.

# 41.3  The Electron's Spin

9. The figure shows a spinning ball of negative charge. Does the magnetic moment of this spinning charge point up, point down, or point to the right? Explain.

10. A bar magnet is moving to the right through a nonuniform magnetic field. The field is weaker toward the bottom of the page and stronger toward the top of the page.

    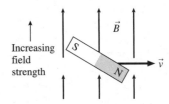

    a. Is there a net force on the magnet? If so, in which direction? Explain.

    b. Will the magnet be deflected by the field? If so, in which direction?

    c. Would the magnet be deflected by a *uniform* magnetic field? Explain why or why not.

11. The figure shows the outcome of a Stern-Gerlach experiment with atoms of element X.

    a. Do the peaks in a Stern-Gerlach experiment represent different values of the atom's total angular momentum or different values of the z-component of its angular momentum? Explain.

    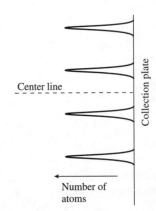

    b. What quantum number characterizes the angular momentum of these atoms? Explain.

# 41.4  Multielectron Atoms

# 41.5  The Periodic Table of the Elements

12. Do the following figures represent a possible electron configuration of an element? If so:
    i.   Identify the element, and
    ii.  Determine if this is the ground state or an excited state.

    If not, why not?

a.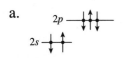

b. $2p$ ———
   $2s$ ⇅⇅
   $1s$ ⇅

c.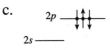

13. Do the following electron configurations represent a possible state of an element? If so:
    i.   Identify the element, and
    ii.  Determine if this is the ground state or an excited state.

    If not, why not?

    a. $1s^2 2s^2 2p^6 3s^2$

    b. $1s^2 2s^2 2p^7 3s$

    c. $1s^2 2s^2 2p^4 3s^2 3p^2$

14. Why is the section of the periodic table labeled as "transition elements" exactly 10 elements wide in all rows?

15. Elements with $Z > 92$ do not occur in nature. They are created in large particle accelerators by smashing together two atoms (such as two lead atoms) and allowing the nuclei to fuse into a larger nucleus. Elements through $Z = 112$ have been discovered this way, and there are as-yet unconfirmed reports of elements beyond $Z = 112$. Suppose that scientists eventually create element 118.

    a. What is the designation of the *last* electron that must be added to create a neutral atom of element 118?

    b. Predict the chemical properties of element 118.

16. What *is* an atom's ionization energy? In other words, if you know the ionization energy of an atom, what is it that you know about the atom?

17. a. According to Figure 41.24, which elements have the highest ionization energies?

    b. Where are these elements located on the periodic table?

    c. What are the basic chemical properties of these elements?

    d. Repeat parts a to c for the elements with the lowest ionization energies.

## 41.6  Excited States and Spectra

18. The figure shows the energy levels of a hypothetical atom.

    a. What is the atom's ionization energy?

    b. In the space below, draw the energy-level diagram as it would appear if the ground state were chosen as the zero of energy. Label each level and the ionization limit with the appropriate energy.

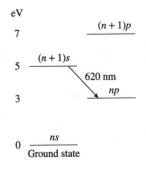

19. The figure shows the energy levels of a hypothetical atom.

    a. What *minimum* kinetic energy (in eV) must an electron have to collisionally excite this atom and cause the emission of a 620 nm photon? Explain.

    b. Can an electron with $K = 6$ eV cause the emission of 620 nm light? If so, what is the final kinetic energy of the electron? If not, why not?

    c. Can a 6 eV photon cause the emission of 620 nm light from this atom? Why or why not?

    d. Can 7 eV photons cause the emission of 620 nm light from these atoms? Why or why not?

20. Seven possible transitions are identified on the energy-level diagram. For each, is this an allowed transition?

    i.   If allowed, is it an emission or an absorption transition?

    ii.  Is the photon infrared, visible, or ultraviolet?

    iii. If not allowed, why not?

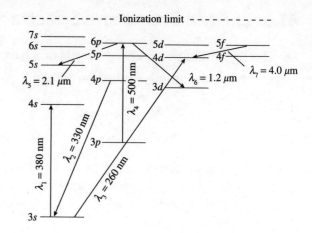

Transition 1:

Transition 2:

Transition 3:

Transition 4:

Transition 5:

Transition 6:

Transition 7:

# 41.7 Lifetimes of Excited States

21. A sample of atoms is excited at $t = 0$ ns. The graph shows the number of excited atoms as a function of time after $t = 0$ ns. Each atom decays back to the ground state by emitting a photon.

    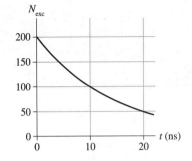

    a. How many atoms were initially excited?

    b. At what time have half of the excited atoms decayed?

    c. How many photons are emitted in the first 20 ns?

    d. At what time will there be 25 atoms remaining in the excited state? Explain.

    e. As best as you can determine from this graph, what is the lifetime of this excited state? Explain how you determined it.

22. At $t = 0$ ns, 1000 atoms are excited to a state that has a lifetime of 5 ns.

    a. Sketch a graph of the number of excited atoms $N_{exc}$ still present at time $t$.

    b. Sketch a graph of the total number of photons $N_{photons}$ that have been emitted by time $t$.

    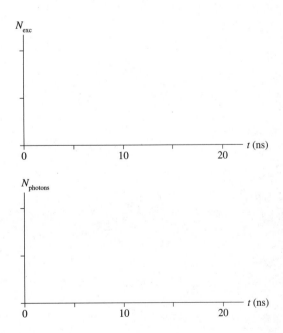

## 41.8 Stimulated Emission and Lasers

23. A photon with energy 2.0 eV is incident on an atom in the
    $p$-state. Does the atom undergo an absorption transition,
    a stimulated emission transition, or neither? Explain.

24. A glass tube contains $2 \times 10^{11}$ atoms, some of which
    are in the ground state and some of which are
    excited. The populations are shown for the atom's
    three energy levels. Is it possible for these atoms to
    be a laser? If so, on which transition would laser
    action occur? If not, why not?

# 42 Nuclear Physics

## 42.1 Nuclear Structure

1. Consider the atoms $^{16}$O, $^{18}$O, $^{18}$F, $^{18}$Ne, and $^{20}$Ne. Some of the questions about these atoms may have more than one answer. Give all answers that apply.

   a. Which atoms are isotopes? _____

   b. Which atoms are isobars? _____

   c. Which atoms have the same chemical properties? _____

   d. Which atoms have the same number of neutrons? _____

   e. Which atoms have the same number of valence electrons? _____
   f. Rank in order, from largest to smallest, the radii of the nuclei of these atoms.

2. Here are several stable or very-long-lived nuclei:

   | | | |
   |---|---|---|
   | $^{58}$Fe | | $^{58}$Ni |
   | $^{60}$Fe | $^{60}$Co | $^{60}$Ni |
   | $^{62}$Fe | | $^{62}$Ni |

   a. Which 2 nuclei are isotopes of $^{60}$Fe? _____

   b. Which 2 nuclei are isobars of $^{60}$Ni? _____

3. The atomic ion S$^-$ has (circle one):

   i.   Lost a proton.
   ii.  Gained an extra electron.

   iii. Either a or b.
   iv.  Turned a proton into a neutron.

   Explain your choice.

4. Given that $m_H = 1.00783$ u, is the mass of a hydrogen atom $^1$H greater than, less than, or equal to $\frac{1}{12}$ the mass of a $^{12}$C atom? Explain.

## 42.2 Nuclear Stability

5. a. Is there a $^{30}$Li ($Z = 3$) nucleus? If so, is it stable or radioactive? If not, why not?

   b. Is there a $^{184}$U ($Z = 92$) nucleus? If so, is it stable or radioactive? If not, why not?

6. a. Is the total binding energy of a nucleus with $A = 200$ more than, less than, or equal to the binding energy of a nucleus with $A = 60$? Explain.

   b. Is a nucleus with $A = 200$ more tightly bound, less tightly bound, or bound equally tightly as a nucleus with $A = 60$? Explain.

7. Rounding slightly, the nucleus $^{3}$He has a binding energy of 2.5 MeV/nucleon and the nucleus $^{6}$Li has a binding energy of 5 MeV/nucleon.

   a. What is the binding energy of $^{3}$He?  _____

   b. What is the binding energy of $^{6}$Li?  _____

   c. Is it energetically possible for two $^{3}$He nuclei to join or fuse together into a $^{6}$Li nucleus? Explain.

   d. Is it energetically possible for a $^{6}$Li nucleus to split or fission into two $^{3}$He nuclei? Explain.

## 42.3  The Strong Force

## 42.4  The Shell Model

8. a. How do we know the strong force exists?

b. How do we know the strong force is short range?

9. For each nuclear energy-level diagram, state whether this represents a nuclear ground state, an excited nuclear state, or an impossible nucleus.

a.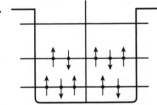

b.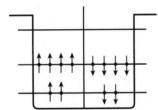

c.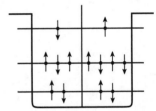

_____          _____          _____

10. Draw energy-level diagrams showing the nucleons in $^6$Li and $^7$Li.

a.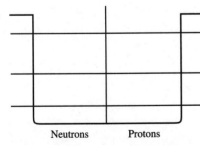

Neutrons    Protons

$^6$Li

b.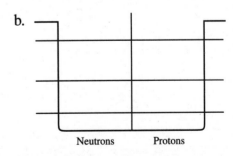

Neutrons    Protons

$^7$Li

## 42.5 Radiation and Radioactivity

11. What is the half-life of this nucleus?

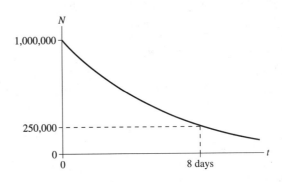

12. A radioactive sample has a half-life of 10 s. 10,000 nuclei are present at $t = 20$ s.

    a. How many nuclei were there at $t = 0$ s?  _____

    b. How many nuclei will there be at $t = 40$ s?  _____

13. Nucleus A decays into nucleus B with a half-life of 10 s. At $t = 0$ s there are 1000 A nuclei and no B nuclei. At what time will there be 750 B nuclei?

14. A radioactive sample with a half-life of 1.0 min is prepared at $t = 0$ min. One particular nucleus has not decayed at $t = 15$ min. What is the probability that this nucleus will decay sometime between $t = 15$ min and $t = 16$ min?

15. You have four radioactive samples:

| Sample | Moles | Activity (Bq) |
| --- | --- | --- |
| A | 1 | 100 |
| B | 10 | 100 |
| C | 100 | 100 |
| D | 100 | 1000 |

    Rank in order, from largest to smallest, the half-lives of these four samples.

    Order:

    Explanation:

## 42.6 Nuclear Decay Mechanisms

16. Identify the unknown X in the following decays:

   a. $^{222}$Rn($Z = 86$) → $^{218}$Po($Z = 84$) + X    X = _____

   b. $^{228}$Ra($Z = 88$) → $^{228}$Ac($Z = 89$) + X    X = _____

   c. $^{140}$Xe($Z = 54$) → $^{140}$Cs($Z = 55$) + X    X = _____

   d. $^{64}$Cu($Z = 29$) → $^{64}$Ni($Z = 28$) + X    X = _____

   e. $^{18}$F($Z = 9$) + X → $^{18}$O($Z = 8$)        X = _____

17. Are the following decays possible? If not, why not?

   a. $^{232}$Th($Z = 90$) → $^{236}$U($Z = 92$) + $\alpha$

   b. $^{238}$Pu($Z = 94$) → $^{236}$U($Z = 92$) + $\alpha$

   c. $^{11}$Na($Z = 11$) → $^{11}$Na($Z = 11$) + $\gamma$

   d. $^{33}$P($Z = 15$) → $^{32}$S($Z = 16$) + $e^-$

18. An alpha particle is ejected from a nucleus with 6 MeV of kinetic energy. Was the alpha particle's kinetic energy inside the nucleus more than, less than, or equal to 6 MeV? Explain.

19. Three alpha particle energy levels in a nuclear potential-energy well are shown below.

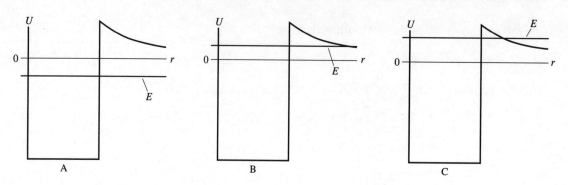

a. Which can undergo alpha decay? _____

b. Of those that decay, which has the longest half-life? _____

c. Of those that decay, which ejects the alpha particle with the largest kinetic energy? _____

20. What kind of decay, if any, can occur for these nuclei?

a. 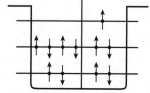   b. 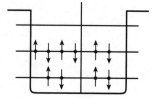   c.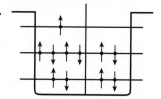

_____   _____   _____

21. Part of the $^{236}$U decay series is shown.
   a. Complete the labeling of atomic numbers and elements on the bottom edge.
   b. Label each arrow to show the type of decay.

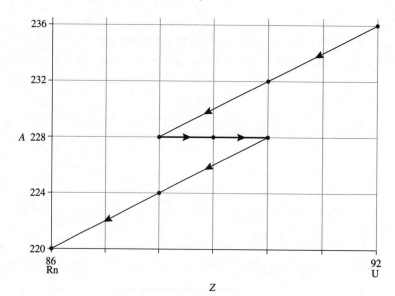

## 42.7  Biological Applications of Nuclear Physics

22. A and B are apples. Apple A is strongly irradiated by nuclear radiation for 1 hour. Apple B is not irradiated. Afterward, in what ways are apples A and B different?

23. Four radiation doses are as follows.

| Dose | Rads | RBE |
|------|------|-----|
| A | 10 | 1 |
| B | 20 | 1 |
| C | 10 | 2 |
| D | 20 | 2 |

   a. Rank in order, from largest to smallest, the amount of energy delivered by these four doses.

   b. Rank in order, from largest to smallest, the biological damage caused by these four doses.

24. Consider the following three decays:

   $^{212}\text{Po} \rightarrow {}^{208}\text{Pb} + \alpha$

   $^{137}\text{Cs} \rightarrow {}^{137}\text{Ba} + e^- + \gamma$

   $^{90}\text{Sr} \rightarrow {}^{90}\text{Y} + e^-$

   Which of the three isotopes, $^{212}\text{Po}$, $^{137}\text{Cs}$, or $^{90}\text{Sr}$, would be most useful as a tracer? Why?

25. The magnetic field through a patient increases from left to right.

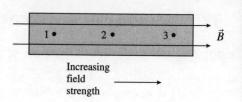

Increasing field strength

a. Are the nmr resonance frequencies at points 1, 2, and 3 the same or different? Explain.

b. If they are different, which is the highest resonance frequency?

# DYNAMICS WORKSHEET  Name _____  Problem _____

## 1) Pictorial Representation
    a. sketch showing important points in the motion
    b. coordinate system
    c. symbols for knowns and unknowns

Known:




Find:

## 2) Physical Representation
    a. motion diagram
    b. force identification
    c. free-body diagram

## 3) Mathematical Representation
    a. start with Newton's first or second law
    b. include other information as needed
    c. solve!

## 4) Assess
    a. units?
    b. reasonable?

# DYNAMICS WORKSHEET   Name _____   Problem _____

## 1) Pictorial Representation
    a. sketch showing important points in the motion
    b. coordinate system
    c. symbols for knowns and unknowns

Known:



Find:

## 2) Physical Representation
    a. motion diagram
    b. force identification
    c. free-body diagram

## 3) Mathematical Representation
    a. start with Newton's first or second law
    b. include other information as needed
    c. solve!

## 4) Assess
    a. units?
    b. reasonable?

# DYNAMICS WORKSHEET   Name _____   Problem _____

## 1) Pictorial Representation
   a. sketch showing important points in the motion
   b. coordinate system
   c. symbols for knowns and unknowns

Known:




Find:

## 2) Physical Representation
   a. motion diagram
   b. force identification
   c. free-body diagram

## 3) Mathematical Representation
   a. start with Newton's first or second law
   b. include other information as needed
   c. solve!

## 4) Assess
   a. units?
   b. reasonable?

# DYNAMICS WORKSHEET  Name _____  Problem _____

## 1) Pictorial Representation
    a. sketch showing important points in the motion
    b. coordinate system
    c. symbols for knowns and unknowns

```
Known:




Find:
```

## 2) Physical Representation
    a. motion diagram
    b. force identification
    c. free-body diagram

## 3) Mathematical Representation
    a. start with Newton's first or second law
    b. include other information as needed
    c. solve!

## 4) Assess
    a. units?
    b. reasonable?

# DYNAMICS WORKSHEET   Name _____   Problem _____

## 1) Pictorial Representation
    a. sketch showing important points in the motion
    b. coordinate system
    c. symbols for knowns and unknowns

Known:



Find:

## 2) Physical Representation
    a. motion diagram
    b. force identification
    c. free-body diagram

## 3) Mathematical Representation
    a. start with Newton's first or second law
    b. include other information as needed
    c. solve!

## 4) Assess
    a. units?
    b. reasonable?

# DYNAMICS WORKSHEET   Name _____   Problem _____

## 1) Pictorial Representation
   a. sketch showing important points in the motion
   b. coordinate system
   c. symbols for knowns and unknowns

Known:



Find:

## 2) Physical Representation
   a. motion diagram
   b. force identification
   c. free-body diagram

## 3) Mathematical Representation
   a. start with Newton's first or second law
   b. include other information as needed
   c. solve!

## 4) Assess
   a. units?
   b. reasonable?

# DYNAMICS WORKSHEET   Name _____ Problem _____

## 1) Pictorial Representation
   a. sketch showing important points in the motion
   b. coordinate system
   c. symbols for knowns and unknowns

+---------------------------+
| Known:                    |
|                           |
|                           |
|                           |
|                           |
|                           |
| Find:                     |
+---------------------------+

## 2) Physical Representation
   a. motion diagram
   b. force identification
   c. free-body diagram

## 3) Mathematical Representation
   a. start with Newton's first or second law
   b. include other information as needed
   c. solve!

## 4) Assess
   a. units?
   b. reasonable?

# DYNAMICS WORKSHEET   Name _____   Problem _____

## 1) Pictorial Representation
    a. sketch showing important points in the motion
    b. coordinate system
    c. symbols for knowns and unknowns

Known:

Find:

## 2) Physical Representation
    a. motion diagram
    b. force identification
    c. free-body diagram

## 3) Mathematical Representation
    a. start with Newton's first or second law
    b. include other information as needed
    c. solve!

## 4) Assess
    a. units?
    b. reasonable?

# DYNAMICS WORKSHEET  Name _____  Problem _____

## 1) Pictorial Representation
    a. sketch showing important points in the motion
    b. coordinate system
    c. symbols for knowns and unknowns

Known:




Find:

## 2) Physical Representation
    a. motion diagram
    b. force identification
    c. free-body diagram

## 3) Mathematical Representation
    a. start with Newton's first or second law
    b. include other information as needed
    c. solve!

## 4) Assess
    a. units?
    b. reasonable?

# DYNAMICS WORKSHEET    Name _____    Problem _____

## 1) Pictorial Representation
   a. sketch showing important points in the motion
   b. coordinate system
   c. symbols for knowns and unknowns

| Known: |
| --- |
| |
| Find: |

## 2) Physical Representation
   a. motion diagram
   b. force identification
   c. free-body diagram

## 3) Mathematical Representation
   a. start with Newton's first or second law
   b. include other information as needed
   c. solve!

## 4) Assess
   a. units?
   b. reasonable?

# DYNAMICS WORKSHEET   Name _____   Problem _____

## 1) Pictorial Representation
   a. sketch showing important points in the motion
   b. coordinate system
   c. symbols for knowns and unknowns

Known:



Find:

## 2) Physical Representation
   a. motion diagram
   b. force identification
   c. free-body diagram

## 3) Mathematical Representation
   a. start with Newton's first or second law
   b. include other information as needed
   c. solve!

## 4) Assess
   a. units?
   b. reasonable?

# DYNAMICS WORKSHEET

Name _____

Problem _____

## 1) Pictorial Representation
    a. sketch showing important points in the motion
    b. coordinate system
    c. symbols for knowns and unknowns

```
Known:




Find:
```

## 2) Physical Representation
    a. motion diagram
    b. force identification
    c. free-body diagram

## 3) Mathematical Representation
    a. start with Newton's first or second law
    b. include other information as needed
    c. solve!

## 4) Assess
    a. units?
    b. reasonable?

# DYNAMICS WORKSHEET  Name _____  Problem _____

## 1) Pictorial Representation
   a. sketch showing important points in the motion
   b. coordinate system
   c. symbols for knowns and unknowns

Known:



Find:

## 2) Physical Representation
   a. motion diagram
   b. force identification
   c. free-body diagram

## 3) Mathematical Representation
   a. start with Newton's first or second law
   b. include other information as needed
   c. solve!

## 4) Assess
   a. units?
   b. reasonable?

# DYNAMICS WORKSHEET  Name _____  Problem _____

## 1) Pictorial Representation
  a. sketch showing important points in the motion
  b. coordinate system
  c. symbols for knowns and unknowns

| Known: |
| --- |
|  |
| Find: |

## 2) Physical Representation
  a. motion diagram
  b. force identification
  c. free-body diagram

## 3) Mathematical Representation
  a. start with Newton's first or second law
  b. include other information as needed
  c. solve!

## 4) Assess
  a. units?
  b. reasonable?

# DYNAMICS WORKSHEET   Name _____   Problem _____

## 1) Pictorial Representation
    a. sketch showing important points in the motion
    b. coordinate system
    c. symbols for knowns and unknowns

Known:



Find:

## 2) Physical Representation
    a. motion diagram
    b. force identification
    c. free-body diagram

## 3) Mathematical Representation
    a. start with Newton's first or second law
    b. include other information as needed
    c. solve!

## 4) Assess
    a. units?
    b. reasonable?

# DYNAMICS WORKSHEET  Name _____  Problem _____

## 1) Pictorial Representation
    a.  sketch showing important points in the motion
    b.  coordinate system
    c.  symbols for knowns and unknowns

Known:

Find:

## 2) Physical Representation
    a.  motion diagram
    b.  force identification
    c.  free-body diagram

## 3) Mathematical Representation
    a.  start with Newton's first or second law
    b.  include other information as needed
    c.  solve!

## 4) Assess
    a.  units?
    b.  reasonable?

# DYNAMICS WORKSHEET    Name _____    Problem _____

## 1) Pictorial Representation
   a. sketch showing important points in the motion
   b. coordinate system
   c. symbols for knowns and unknowns

| Known: |
| --- |
| |
| Find: |

## 2) Physical Representation
   a. motion diagram
   b. force identification
   c. free-body diagram

## 3) Mathematical Representation
   a. start with Newton's first or second law
   b. include other information as needed
   c. solve!

## 4) Assess
   a. units?
   b. reasonable?

# DYNAMICS WORKSHEET  Name _____  Problem _____

## 1) Pictorial Representation
    a. sketch showing important points in the motion
    b. coordinate system
    c. symbols for knowns and unknowns

Known:



Find:

## 2) Physical Representation
    a. motion diagram
    b. force identification
    c. free-body diagram

## 3) Mathematical Representation
    a. start with Newton's first or second law
    b. include other information as needed
    c. solve!

## 4) Assess
    a. units?
    b. reasonable?

# DYNAMICS WORKSHEET   Name _____   Problem _____

## 1) Pictorial Representation
    a. sketch showing important points in the motion
    b. coordinate system
    c. symbols for knowns and unknowns

Known:




Find:

## 2) Physical Representation
    a. motion diagram
    b. force identification
    c. free-body diagram

## 3) Mathematical Representation
    a. start with Newton's first or second law
    b. include other information as needed
    c. solve!

## 4) Assess
    a. units?
    b. reasonable?

# DYNAMICS WORKSHEET   Name _____   Problem _____

## 1) Pictorial Representation
   a. sketch showing important points in the motion
   b. coordinate system
   c. symbols for knowns and unknowns

Known:



Find:

## 2) Physical Representation
   a. motion diagram
   b. force identification
   c. free-body diagram

## 3) Mathematical Representation
   a. start with Newton's first or second law
   b. include other information as needed
   c. solve!

## 4) Assess
   a. units?
   b. reasonable?

# DYNAMICS WORKSHEET    Name _____    Problem _____

## 1) Pictorial Representation
    a. sketch showing important points in the motion
    b. coordinate system
    c. symbols for knowns and unknowns

| Known: |
| --- |
| |
| Find: |

## 2) Physical Representation
    a. motion diagram
    b. force identification
    c. free-body diagram

## 3) Mathematical Representation
    a. start with Newton's first or second law
    b. include other information as needed
    c. solve!

## 4) Assess
    a. units?
    b. reasonable?

# DYNAMICS WORKSHEET   Name _____   Problem _____

## 1) Pictorial Representation
    a. sketch showing important points in the motion
    b. coordinate system
    c. symbols for knowns and unknowns

Known:



Find:

## 2) Physical Representation
    a. motion diagram
    b. force identification
    c. free-body diagram

## 3) Mathematical Representation
    a. start with Newton's first or second law
    b. include other information as needed
    c. solve!

## 4) Assess
    a. units?
    b. reasonable?

# DYNAMICS WORKSHEET    Name _____    Problem _____

## 1) Pictorial Representation
   a. sketch showing important points in the motion
   b. coordinate system
   c. symbols for knowns and unknowns

```
Known:







Find:
```

## 2) Physical Representation
   a. motion diagram
   b. force identification
   c. free-body diagram

## 3) Mathematical Representation
   a. start with Newton's first or second law
   b. include other information as needed
   c. solve!

## 4) Assess
   a. units?
   b. reasonable?

# DYNAMICS WORKSHEET    Name _____    Problem _____

## 1) Pictorial Representation
   a. sketch showing important points in the motion
   b. coordinate system
   c. symbols for knowns and unknowns

Known:



Find:

## 2) Physical Representation
   a. motion diagram
   b. force identification
   c. free-body diagram

## 3) Mathematical Representation
   a. start with Newton's first or second law
   b. include other information as needed
   c. solve!

## 4) Assess
   a. units?
   b. reasonable?

# DYNAMICS WORKSHEET    Name _____    Problem _____

## 1) Pictorial Representation
     a. sketch showing important points in the motion
     b. coordinate system
     c. symbols for knowns and unknowns

Known:




Find:

## 2) Physical Representation
     a. motion diagram
     b. force identification
     c. free-body diagram

## 3) Mathematical Representation
     a. start with Newton's first or second law
     b. include other information as needed
     c. solve!

## 4) Assess
     a. units?
     b. reasonable?

# DYNAMICS WORKSHEET  Name _____    Problem _____

## 1) Pictorial Representation
   a. sketch showing important points in the motion
   b. coordinate system
   c. symbols for knowns and unknowns

Known:



Find:

## 2) Physical Representation
   a. motion diagram
   b. force identification
   c. free-body diagram

## 3) Mathematical Representation
   a. start with Newton's first or second law
   b. include other information as needed
   c. solve!

## 4) Assess
   a. units?
   b. reasonable?

# DYNAMICS WORKSHEET   Name _____   Problem _____

## 1) Pictorial Representation
   a. sketch showing important points in the motion
   b. coordinate system
   c. symbols for knowns and unknowns

| Known: |
| --- |
| |
| Find: |

## 2) Physical Representation
   a. motion diagram
   b. force identification
   c. free-body diagram

## 3) Mathematical Representation
   a. start with Newton's first or second law
   b. include other information as needed
   c. solve!

## 4) Assess
   a. units?
   b. reasonable?

# DYNAMICS WORKSHEET   Name _____   Problem _____

## 1) Pictorial Representation
   a. sketch showing important points in the motion
   b. coordinate system
   c. symbols for knowns and unknowns

Known:



Find:

## 2) Physical Representation
   a. motion diagram
   b. force identification
   c. free-body diagram

## 3) Mathematical Representation
   a. start with Newton's first or second law
   b. include other information as needed
   c. solve!

## 4) Assess
   a. units?
   b. reasonable?

# DYNAMICS WORKSHEET  Name _____  Problem _____

## 1) Pictorial Representation
   a. sketch showing important points in the motion
   b. coordinate system
   c. symbols for knowns and unknowns

```
┌─────────────────────────────────┐
│ Known:                          │
│                                 │
│                                 │
│                                 │
│                                 │
│                                 │
├─────────────────────────────────┤
│ Find:                           │
└─────────────────────────────────┘
```

## 2) Physical Representation
   a. motion diagram
   b. force identification
   c. free-body diagram

## 3) Mathematical Representation
   a. start with Newton's first or second law
   b. include other information as needed
   c. solve!

## 4) Assess
   a. units?
   b. reasonable?

# DYNAMICS WORKSHEET   Name _____   Problem _____

## 1) Pictorial Representation
    a. sketch showing important points in the motion
    b. coordinate system
    c. symbols for knowns and unknowns

> Known:
>
>
>
>
>
>
>
>
> Find:

## 2) Physical Representation
    a. motion diagram
    b. force identification
    c. free-body diagram

## 3) Mathematical Representation
    a. start with Newton's first or second law
    b. include other information as needed
    c. solve!

## 4) Assess
    a. units?
    b. reasonable?

# DYNAMICS WORKSHEET  Name _____  Problem _____

## 1) Pictorial Representation
   a. sketch showing important points in the motion
   b. coordinate system
   c. symbols for knowns and unknowns

| Known: |
| --- |
| |
| Find: |

## 2) Physical Representation
   a. motion diagram
   b. force identification
   c. free-body diagram

## 3) Mathematical Representation
   a. start with Newton's first or second law
   b. include other information as needed
   c. solve!

## 4) Assess
   a. units?
   b. reasonable?

# DYNAMICS WORKSHEET   Name _____   Problem _____

## 1) Pictorial Representation
    a. sketch showing important points in the motion
    b. coordinate system
    c. symbols for knowns and unknowns

Known:



Find:

## 2) Physical Representation
    a. motion diagram
    b. force identification
    c. free-body diagram

## 3) Mathematical Representation
    a. start with Newton's first or second law
    b. include other information as needed
    c. solve!

## 4) Assess
    a. units?
    b. reasonable?

# DYNAMICS WORKSHEET  Name _____  Problem _____

## 1) Pictorial Representation
    a. sketch showing important points in the motion
    b. coordinate system
    c. symbols for knowns and unknowns

Known:

Find:

## 2) Physical Representation
    a. motion diagram
    b. force identification
    c. free-body diagram

## 3) Mathematical Representation
    a. start with Newton's first or second law
    b. include other information as needed
    c. solve!

## 4) Assess
    a. units?
    b. reasonable?

# DYNAMICS WORKSHEET   Name _____   Problem _____

## 1) Pictorial Representation
    a. sketch showing important points in the motion
    b. coordinate system
    c. symbols for knowns and unknowns

Known:



Find:

## 2) Physical Representation
    a. motion diagram
    b. force identification
    c. free-body diagram

## 3) Mathematical Representation
    a. start with Newton's first or second law
    b. include other information as needed
    c. solve!

## 4) Assess
    a. units?
    b. reasonable?

# DYNAMICS WORKSHEET   Name _____   Problem _____

## 1) Pictorial Representation
    a. sketch showing important points in the motion
    b. coordinate system
    c. symbols for knowns and unknowns

Known:



Find:

## 2) Physical Representation
    a. motion diagram
    b. force identification
    c. free-body diagram

## 3) Mathematical Representation
    a. start with Newton's first or second law
    b. include other information as needed
    c. solve!

## 4) Assess
    a. units?
    b. reasonable?

# DYNAMICS WORKSHEET    Name _____    Problem _____

## 1) Pictorial Representation
   a. sketch showing important points in the motion
   b. coordinate system
   c. symbols for knowns and unknowns

| Known: |
| --- |
| |
| Find: |

## 2) Physical Representation
   a. motion diagram
   b. force identification
   c. free-body diagram

## 3) Mathematical Representation
   a. start with Newton's first or second law
   b. include other information as needed
   c. solve!

## 4) Assess
   a. units?
   b. reasonable?

# DYNAMICS WORKSHEET   Name _____   Problem _____

## 1) Pictorial Representation
   a. sketch showing important points in the motion
   b. coordinate system
   c. symbols for knowns and unknowns

| Known: |
| --- |
| |
| Find: |

## 2) Physical Representation
   a. motion diagram
   b. force identification
   c. free-body diagram

## 3) Mathematical Representation
   a. start with Newton's first or second law
   b. include other information as needed
   c. solve!

## 4) Assess
   a. units?
   b. reasonable?

# DYNAMICS WORKSHEET   Name _____   Problem _____

## 1) Pictorial Representation
   a. sketch showing important points in the motion
   b. coordinate system
   c. symbols for knowns and unknowns

Known:


Find:

## 2) Physical Representation
   a. motion diagram
   b. force identification
   c. free-body diagram

## 3) Mathematical Representation
   a. start with Newton's first or second law
   b. include other information as needed
   c. solve!

## 4) Assess
   a. units?
   b. reasonable?

# DYNAMICS WORKSHEET   Name _____   Problem _____

## 1) Pictorial Representation
   a. sketch showing important points in the motion
   b. coordinate system
   c. symbols for knowns and unknowns

```
Known:




Find:
```

## 2) Physical Representation
   a. motion diagram
   b. force identification
   c. free-body diagram

## 3) Mathematical Representation
   a. start with Newton's first or second law
   b. include other information as needed
   c. solve!

## 4) Assess
   a. units?
   b. reasonable?

**DYNAMICS WORKSHEET**  Name _____  Problem _____

## 1) Pictorial Representation

    a. sketch showing important points in the motion

    b. coordinate system

    c. symbols for knowns and unknowns

Known:




Find:

## 2) Physical Representation

    a. motion diagram

    b. force identification

    c. free-body diagram

## 3) Mathematical Representation

    a. start with Newton's first or second law

    b. include other information as needed

    c. solve!

## 4) Assess

    a. units?

    b. reasonable?

# DYNAMICS WORKSHEET   Name _____   Problem _____

## 1) Pictorial Representation
    a. sketch showing important points in the motion
    b. coordinate system
    c. symbols for knowns and unknowns

Known:



Find:

## 2) Physical Representation
    a. motion diagram
    b. force identification
    c. free-body diagram

## 3) Mathematical Representation
    a. start with Newton's first or second law
    b. include other information as needed
    c. solve!

## 4) Assess
    a. units?
    b. reasonable?

# DYNAMICS WORKSHEET   Name _____ Problem _____

## 1) Pictorial Representation
   a. sketch showing important points in the motion
   b. coordinate system
   c. symbols for knowns and unknowns

> Known:
>
>
>
>
>
>
>
> Find:

## 2) Physical Representation
   a. motion diagram
   b. force identification
   c. free-body diagram

## 3) Mathematical Representation
   a. start with Newton's first or second law
   b. include other information as needed
   c. solve!

## 4) Assess
   a. units?
   b. reasonable?

# DYNAMICS WORKSHEET   Name _____   Problem _____

## 1) Pictorial Representation
   a. sketch showing important points in the motion
   b. coordinate system
   c. symbols for knowns and unknowns

```
┌─────────────────────────────────┐
│ Known:                          │
│                                 │
│                                 │
│                                 │
│                                 │
│                                 │
├─────────────────────────────────┤
│ Find:                           │
└─────────────────────────────────┘
```

## 2) Physical Representation
   a. motion diagram
   b. force identification
   c. free-body diagram

## 3) Mathematical Representation
   a. start with Newton's first or second law
   b. include other information as needed
   c. solve!

## 4) Assess
   a. units?
   b. reasonable?

# DYNAMICS WORKSHEET  Name _____    Problem _____

## 1) Pictorial Representation
   a. sketch showing important points in the motion
   b. coordinate system
   c. symbols for knowns and unknowns

Known:



Find:

## 2) Physical Representation
   a. motion diagram
   b. force identification
   c. free-body diagram

## 3) Mathematical Representation
   a. start with Newton's first or second law
   b. include other information as needed
   c. solve!

## 4) Assess
   a. units?
   b. reasonable?

# DYNAMICS WORKSHEET   Name _____   Problem _____

## 1) Pictorial Representation
    a. sketch showing important points in the motion
    b. coordinate system
    c. symbols for knowns and unknowns

Known:




Find:

## 2) Physical Representation
    a. motion diagram
    b. force identification
    c. free-body diagram

## 3) Mathematical Representation
    a. start with Newton's first or second law
    b. include other information as needed
    c. solve!

## 4) Assess
    a. units?
    b. reasonable?

# DYNAMICS WORKSHEET   Name _____   Problem _____

## 1) Pictorial Representation
    a. sketch showing important points in the motion
    b. coordinate system
    c. symbols for knowns and unknowns

Known:



Find:

## 2) Physical Representation
    a. motion diagram
    b. force identification
    c. free-body diagram

## 3) Mathematical Representation
    a. start with Newton's first or second law
    b. include other information as needed
    c. solve!

## 4) Assess
    a. units?
    b. reasonable?

# DYNAMICS WORKSHEET   Name _____   Problem _____

## 1) Pictorial Representation
    a. sketch showing important points in the motion
    b. coordinate system
    c. symbols for knowns and unknowns

Known:

Find:

## 2) Physical Representation
    a. motion diagram
    b. force identification
    c. free-body diagram

## 3) Mathematical Representation
    a. start with Newton's first or second law
    b. include other information as needed
    c. solve!

## 4) Assess
    a. units?
    b. reasonable?

# DYNAMICS WORKSHEET   Name _____   Problem _____

## 1) Pictorial Representation
    a. sketch showing important points in the motion
    b. coordinate system
    c. symbols for knowns and unknowns

Known:



Find:

## 2) Physical Representation
    a. motion diagram
    b. force identification
    c. free-body diagram

## 3) Mathematical Representation
    a. start with Newton's first or second law
    b. include other information as needed
    c. solve!

## 4) Assess
    a. units?
    b. reasonable?

# DYNAMICS WORKSHEET  Name _____  Problem _____

## 1) Pictorial Representation
   a. sketch showing important points in the motion
   b. coordinate system
   c. symbols for knowns and unknowns

```
┌─────────────────────────────────┐
│ Known:                          │
│                                 │
│                                 │
│                                 │
│                                 │
│                                 │
├─────────────────────────────────┤
│ Find:                           │
└─────────────────────────────────┘
```

## 2) Physical Representation
   a. motion diagram
   b. force identification
   c. free-body diagram

## 3) Mathematical Representation
   a. start with Newton's first or second law
   b. include other information as needed
   c. solve!

## 4) Assess
   a. units?
   b. reasonable?

# DYNAMICS WORKSHEET  Name _____  Problem _____

## 1) Pictorial Representation
    a. sketch showing important points in the motion
    b. coordinate system
    c. symbols for knowns and unknowns

Known:



Find:

## 2) Physical Representation
    a. motion diagram
    b. force identification
    c. free-body diagram

## 3) Mathematical Representation
    a. start with Newton's first or second law
    b. include other information as needed
    c. solve!

## 4) Assess
    a. units?
    b. reasonable?

# DYNAMICS WORKSHEET   Name _____   Problem _____

## 1) Pictorial Representation
   a. sketch showing important points in the motion
   b. coordinate system
   c. symbols for knowns and unknowns

Known:



Find:

## 2) Physical Representation
   a. motion diagram
   b. force identification
   c. free-body diagram

## 3) Mathematical Representation
   a. start with Newton's first or second law
   b. include other information as needed
   c. solve!

## 4) Assess
   a. units?
   b. reasonable?

# DYNAMICS WORKSHEET   Name _____   Problem _____

## 1) Pictorial Representation
   a. sketch showing important points in the motion
   b. coordinate system
   c. symbols for knowns and unknowns

| Known: |
| --- |
| Find: |

## 2) Physical Representation
   a. motion diagram
   b. force identification
   c. free-body diagram

## 3) Mathematical Representation
   a. start with Newton's first or second law
   b. include other information as needed
   c. solve!

## 4) Assess
   a. units?
   b. reasonable?

# DYNAMICS WORKSHEET   Name _____   Problem _____

## 1) Pictorial Representation
  a. sketch showing important points in the motion
  b. coordinate system
  c. symbols for knowns and unknowns

| Known: |
| --- |
| |
| Find: |

## 2) Physical Representation
  a. motion diagram
  b. force identification
  c. free-body diagram

## 3) Mathematical Representation
  a. start with Newton's first or second law
  b. include other information as needed
  c. solve!

## 4) Assess
  a. units?
  b. reasonable?

# DYNAMICS WORKSHEET   Name _____   Problem _____

## 1) Pictorial Representation
    a. sketch showing important points in the motion
    b. coordinate system
    c. symbols for knowns and unknowns

Known:

Find:

## 2) Physical Representation
    a. motion diagram
    b. force identification
    c. free-body diagram

## 3) Mathematical Representation
    a. start with Newton's first or second law
    b. include other information as needed
    c. solve!

## 4) Assess
    a. units?
    b. reasonable?

# DYNAMICS WORKSHEET    Name _____    Problem _____

## 1) Pictorial Representation
    a. sketch showing important points in the motion
    b. coordinate system
    c. symbols for knowns and unknowns

Known:



Find:

## 2) Physical Representation
    a. motion diagram
    b. force identification
    c. free-body diagram

## 3) Mathematical Representation
    a. start with Newton's first or second law
    b. include other information as needed
    c. solve!

## 4) Assess
    a. units?
    b. reasonable?

# DYNAMICS WORKSHEET   Name _____   Problem _____

## 1) Pictorial Representation
    a. sketch showing important points in the motion
    b. coordinate system
    c. symbols for knowns and unknowns

> Known:
>
>
>
>
>
>
>
> Find:

## 2) Physical Representation
    a. motion diagram
    b. force identification
    c. free-body diagram

## 3) Mathematical Representation
    a. start with Newton's first or second law
    b. include other information as needed
    c. solve!

## 4) Assess
    a. units?
    b. reasonable?

# DYNAMICS WORKSHEET  Name _____ Problem _____

## 1) Pictorial Representation
   a. sketch showing important points in the motion
   b. coordinate system
   c. symbols for knowns and unknowns

| Known: |
| --- |
|  |
| Find: |

## 2) Physical Representation
   a. motion diagram
   b. force identification
   c. free-body diagram

## 3) Mathematical Representation
   a. start with Newton's first or second law
   b. include other information as needed
   c. solve!

## 4) Assess
   a. units?
   b. reasonable?

# DYNAMICS WORKSHEET   Name _____   Problem _____

## 1) Pictorial Representation
   a. sketch showing important points in the motion
   b. coordinate system
   c. symbols for knowns and unknowns

| Known: |
| --- |
| |
| Find: |

## 2) Physical Representation
   a. motion diagram
   b. force identification
   c. free-body diagram

## 3) Mathematical Representation
   a. start with Newton's first or second law
   b. include other information as needed
   c. solve!

## 4) Assess
   a. units?
   b. reasonable?

# DYNAMICS WORKSHEET   Name _____   Problem _____

## 1) Pictorial Representation
   a. sketch showing important points in the motion
   b. coordinate system
   c. symbols for knowns and unknowns

Known:

Find:

## 2) Physical Representation
   a. motion diagram
   b. force identification
   c. free-body diagram

## 3) Mathematical Representation
   a. start with Newton's first or second law
   b. include other information as needed
   c. solve!

## 4) Assess
   a. units?
   b. reasonable?

# DYNAMICS WORKSHEET   Name _____ Problem _____

## 1) Pictorial Representation
    a. sketch showing important points in the motion
    b. coordinate system
    c. symbols for knowns and unknowns

Known:



Find:

## 2) Physical Representation
    a. motion diagram
    b. force identification
    c. free-body diagram

## 3) Mathematical Representation
    a. start with Newton's first or second law
    b. include other information as needed
    c. solve!

## 4) Assess
    a. units?
    b. reasonable?

# DYNAMICS WORKSHEET    Name _____    Problem _____

## 1) Pictorial Representation
   a. sketch showing important points in the motion
   b. coordinate system
   c. symbols for knowns and unknowns

| Known: |
| --- |
|  |
| Find: |

## 2) Physical Representation
   a. motion diagram
   b. force identification
   c. free-body diagram

## 3) Mathematical Representation
   a. start with Newton's first or second law
   b. include other information as needed
   c. solve!

## 4) Assess
   a. units?
   b. reasonable?

# DYNAMICS WORKSHEET  Name _____  Problem _____

## 1) Pictorial Representation
    a.  sketch showing important points in the motion
    b.  coordinate system
    c.  symbols for knowns and unknowns

Known:

Find:

## 2) Physical Representation
    a.  motion diagram
    b.  force identification
    c.  free-body diagram

## 3) Mathematical Representation
    a.  start with Newton's first or second law
    b.  include other information as needed
    c.  solve!

## 4) Assess
    a.  units?
    b.  reasonable?

# MOMENTUM WORKSHEET

Name _____ Problem _____

## 1) Pictorial Representation
   a. sketch of before and after
   b. coordinate system
   c. symbols for knowns and unknowns

Known:



Find:

## 2) Momentum Bar Chart

What is the system?

Is this an isolated system during all or part of the problem?

What forces, if any, exert impulses on the system?

$$+$$

$$0 \quad \underline{\quad} \; + \; \underline{\quad} \; = \; \underline{\quad}$$

$$-$$

$$P_{ix} \quad + \quad J_x \quad = \quad P_{fx}$$

## 3) Mathematical Representation
   a. momentum conservation or impulse/momentum
   b. use Newton's law or kinematics as needed
   c. solve!

## 4) Assess
   a. units?
   b. reasonable?

# MOMENTUM WORKSHEET   Name _____   Problem _____

## 1) Pictorial Representation
   a.  sketch of before and after
   b.  coordinate system
   c.  symbols for knowns and unknowns

Known:



Find:

## 2) Momentum Bar Chart

What is the system?

Is this an isolated system during all or part of the problem?

What forces, if any, exert impulses on the system?

$$P_{ix} \quad + \quad J_x \quad = \quad P_{fx}$$

## 3) Mathematical Representation
   a.  momentum conservation or impulse/momentum
   b.  use Newton's law or kinematics as needed
   c.  solve!

## 4) Assess
   a.  units?
   b.  reasonable?

# MOMENTUM WORKSHEET   Name _____   Problem _____

## 1) Pictorial Representation
   a. sketch of before and after
   b. coordinate system
   c. symbols for knowns and unknowns

Known:



Find:

## 2) Momentum Bar Chart

What is the system?

Is this an isolated system during all or part of the problem?

What forces, if any, exert impulses on the system?

$$P_{ix} \quad + \quad J_x \quad = \quad P_{fx}$$

## 3) Mathematical Representation
   a. momentum conservation or impulse/momentum
   b. use Newton's law or kinematics as needed
   c. solve!

## 4) Assess
   a. units?
   b. reasonable?

# MOMENTUM WORKSHEET  Name _____  Problem _____

## 1) Pictorial Representation
    a. sketch of before and after
    b. coordinate system
    c. symbols for knowns and unknowns

Known:



Find:

## 2) Momentum Bar Chart

What is the system?

Is this an isolated system during all or part of the problem?

What forces, if any, exert impulses on the system?

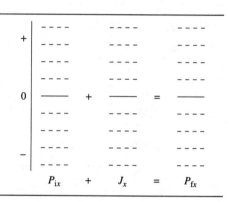

$$P_{ix} \quad + \quad J_x \quad = \quad P_{fx}$$

## 3) Mathematical Representation
    a. momentum conservation or impulse/momentum
    b. use Newton's law or kinematics as needed
    c. solve!

## 4) Assess
    a. units?
    b. reasonable?

# MOMENTUM WORKSHEET    Name _____    Problem _____

## 1) Pictorial Representation
    a.  sketch of before and after
    b.  coordinate system
    c.  symbols for knowns and unknowns

Known:



Find:

## 2) Momentum Bar Chart

    What is the system?

    Is this an isolated system during all or part of the problem?

    What forces, if any, exert impulses on the system?

$$P_{ix} \quad + \quad J_x \quad = \quad P_{fx}$$

## 3) Mathematical Representation
    a.  momentum conservation or impulse/momentum
    b.  use Newton's law or kinematics as needed
    c.  solve!

## 4) Assess
    a.  units?
    b.  reasonable?

# MOMENTUM WORKSHEET   Name _____   Problem _____

## 1) Pictorial Representation
   a. sketch of before and after
   b. coordinate system
   c. symbols for knowns and unknowns

| Known: |
| --- |
| |
| Find: |

## 2) Momentum Bar Chart

What is the system?

Is this an isolated system during all or part of the problem?

What forces, if any, exert impulses on the system?

$$+$$

$$0$$

$$-$$

$$P_{ix} \quad + \quad J_x \quad = \quad P_{fx}$$

## 3) Mathematical Representation
   a. momentum conservation or impulse/momentum
   b. use Newton's law or kinematics as needed
   c. solve!

## 4) Assess
   a. units?
   b. reasonable?

# MOMENTUM WORKSHEET   Name _____   Problem _____

## 1) Pictorial Representation
    a. sketch of before and after
    b. coordinate system
    c. symbols for knowns and unknowns

Known:




Find:

## 2) Momentum Bar Chart

What is the system?


Is this an isolated system during all or part of the problem?


What forces, if any, exert impulses on the system?

$$P_{ix} \quad + \quad J_x \quad = \quad P_{fx}$$

## 3) Mathematical Representation
    a. momentum conservation or impulse/momentum
    b. use Newton's law or kinematics as needed
    c. solve!

## 4) Assess
    a. units?
    b. reasonable?

# MOMENTUM WORKSHEET   Name _____   Problem _____

## 1) Pictorial Representation
    a. sketch of before and after
    b. coordinate system
    c. symbols for knowns and unknowns

Known:



Find:

## 2) Momentum Bar Chart
    What is the system?

    Is this an isolated system during all or part of the problem?

    What forces, if any, exert impulses on the system?

$$P_{ix} \;+\; J_x \;=\; P_{fx}$$

## 3) Mathematical Representation
    a. momentum conservation or impulse/momentum
    b. use Newton's law or kinematics as needed
    c. solve!

## 4) Assess
    a. units?
    b. reasonable?

# MOMENTUM WORKSHEET    Name _____    Problem _____

## 1) Pictorial Representation
   a. sketch of before and after
   b. coordinate system
   c. symbols for knowns and unknowns

| Known: |
| --- |
|  |
| Find: |

## 2) Momentum Bar Chart

What is the system?

Is this an isolated system during all or part of the problem?

What forces, if any, exert impulses on the system?

$$+$$

$$0$$

$$-$$

$$P_{ix} \quad + \quad J_x \quad = \quad P_{fx}$$

## 3) Mathematical Representation
   a. momentum conservation or impulse/momentum
   b. use Newton's law or kinematics as needed
   c. solve!

## 4) Assess
   a. units?
   b. reasonable?

# MOMENTUM WORKSHEET   Name _____   Problem _____

## 1) Pictorial Representation
   a. sketch of before and after
   b. coordinate system
   c. symbols for knowns and unknowns

Known:

Find:

## 2) Momentum Bar Chart

What is the system?

Is this an isolated system during all or part of the problem?

What forces, if any, exert impulses on the system?

$$+ \qquad 0 \qquad -$$

$$P_{ix} \quad + \quad J_x \quad = \quad P_{fx}$$

## 3) Mathematical Representation
   a. momentum conservation or impulse/momentum
   b. use Newton's law or kinematics as needed
   c. solve!

## 4) Assess
   a. units?
   b. reasonable?

# MOMENTUM WORKSHEET   Name _____   Problem _____

## 1) Pictorial Representation
   a. sketch of before and after
   b. coordinate system
   c. symbols for knowns and unknowns

Known:



Find:

## 2) Momentum Bar Chart

What is the system?


Is this an isolated system during all or part of the problem?


What forces, if any, exert impulses on the system?

$$+ \qquad 0 \qquad -$$

$$P_{ix} \quad + \quad J_x \quad = \quad P_{fx}$$

## 3) Mathematical Representation
   a. momentum conservation or impulse/momentum
   b. use Newton's law or kinematics as needed
   c. solve!

## 4) Assess
   a. units?
   b. reasonable?

# MOMENTUM WORKSHEET   Name _____   Problem _____

## 1) Pictorial Representation
   a. sketch of before and after
   b. coordinate system
   c. symbols for knowns and unknowns

Known:



Find:

## 2) Momentum Bar Chart

What is the system?

Is this an isolated system during all or part of the problem?

What forces, if any, exert impulses on the system?

$$P_{ix} + J_x = P_{fx}$$

## 3) Mathematical Representation
   a. momentum conservation or impulse/momentum
   b. use Newton's law or kinematics as needed
   c. solve!

## 4) Assess
   a. units?
   b. reasonable?

# ENERGY WORKSHEET   Name _____   Problem _____

## 1) Pictorial Representation
   a. sketch of before and after
   b. coordinate system
   c. symbols for knowns and unknowns

| Known: |
| --- |
| |
| Find: |

## 2) Energy Bar Chart

What is the system? _____

Potential energies? _____

Nonconservative forces? _____

External forces? _____

Is mechanical energy conserved? _____

$$K_i \;+\; U_i \;+\; W_{ext} \;=\; K_f \;+\; U_f \;+\; \Delta E_{th}$$

## 3) Mathematical Representation
   a. energy conservation or work/energy
   b. use Newton's law or momentum as needed
   c. solve!

## 4) Assess
   a. units?
   b. reasonable?

# ENERGY WORKSHEET   Name _____   Problem _____

## 1) Pictorial Representation
   a. sketch of before and after
   b. coordinate system
   c. symbols for knowns and unknowns

Known:




Find:

## 2) Energy Bar Chart

What is the system? _____

Potential energies? _____

Nonconservative forces? _____

External forces? _____

Is mechanical energy conserved? _____

$$K_i \quad + \quad U_i \quad + \quad W_{ext} \quad = \quad K_f \quad + \quad U_f \quad + \quad \Delta E_{th}$$

## 3) Mathematical Representation
   a. energy conservation or work/energy
   b. use Newton's law or momentum as needed
   c. solve!

## 4) Assess
   a. units?
   b. reasonable?

# ENERGY WORKSHEET   Name _____   Problem _____

## 1) Pictorial Representation
   a. sketch of before and after
   b. coordinate system
   c. symbols for knowns and unknowns

Known:



Find:

## 2) Energy Bar Chart

What is the system? _____

Potential energies? _____

Nonconservative forces? _____

External forces? _____

Is mechanical energy conserved? _____

$$K_i \quad + \quad U_i \quad + \quad W_{ext} \quad = \quad K_f \quad + \quad U_f \quad + \quad \Delta E_{th}$$

## 3) Mathematical Representation
   a. energy conservation or work/energy
   b. use Newton's law or momentum as needed
   c. solve!

## 4) Assess
   a. units?
   b. reasonable?

# ENERGY WORKSHEET   Name _____   Problem _____

## 1) Pictorial Representation
a. sketch of before and after
b. coordinate system
c. symbols for knowns and unknowns

Known:



Find:

## 2) Energy Bar Chart

What is the system? _____

Potential energies? _____

Nonconservative forces? _____

External forces? _____

Is mechanical energy conserved? _____

$$K_i \;+\; U_i \;+\; W_{ext} \;=\; K_f \;+\; U_f \;+\; \Delta E_{th}$$

## 3) Mathematical Representation
a. energy conservation or work/energy
b. use Newton's law or momentum as needed
c. solve!

## 4) Assess
a. units?
b. reasonable?

# ENERGY WORKSHEET   Name _____   Problem _____

## 1) Pictorial Representation
   a. sketch of before and after
   b. coordinate system
   c. symbols for knowns and unknowns

Known:



Find:

## 2) Energy Bar Chart

What is the system? _____

Potential energies? _____

Nonconservative forces? _____

External forces? _____

Is mechanical energy conserved? _____

$$K_i \quad + \quad U_i \quad + \quad W_{ext} \quad = \quad K_f \quad + \quad U_f \quad + \quad \Delta E_{th}$$

## 3) Mathematical Representation
   a. energy conservation or work/energy
   b. use Newton's law or momentum as needed
   c. solve!

## 4) Assess
   a. units?
   b. reasonable?

# ENERGY WORKSHEET   Name _____   Problem _____

## 1) Pictorial Representation
    a. sketch of before and after
    b. coordinate system
    c. symbols for knowns and unknowns

Known:



Find:

## 2) Energy Bar Chart

What is the system? _____

Potential energies? _____

Nonconservative forces? _____

External forces? _____

Is mechanical energy conserved? _____

$$K_i \quad + \quad U_i \quad + \quad W_{ext} \quad = \quad K_f \quad + \quad U_f \quad + \quad \Delta E_{th}$$

## 3) Mathematical Representation
    a. energy conservation or work/energy
    b. use Newton's law or momentum as needed
    c. solve!

## 4) Assess
    a. units?
    b. reasonable?

# ENERGY WORKSHEET   Name _____   Problem _____

## 1) Pictorial Representation
   a. sketch of before and after
   b. coordinate system
   c. symbols for knowns and unknowns

Known:



Find:

## 2) Energy Bar Chart

What is the system? _____

Potential energies? _____

Nonconservative forces? _____

External forces? _____

Is mechanical energy conserved? _____

$$K_i \quad + \quad U_i \quad + \quad W_{ext} \quad = \quad K_f \quad + \quad U_f \quad + \quad \Delta E_{th}$$

## 3) Mathematical Representation
   a. energy conservation or work/energy
   b. use Newton's law or momentum as needed
   c. solve!

## 4) Assess
   a. units?
   b. reasonable?

# ENERGY WORKSHEET   Name _____   Problem _____

## 1) Pictorial Representation
   a.  sketch of before and after
   b.  coordinate system
   c.  symbols for knowns and unknowns

Known:



Find:

## 2) Energy Bar Chart

What is the system? _____

Potential energies? _____

Nonconservative forces? _____

External forces? _____

Is mechanical energy conserved? _____

$$K_i \; + \; U_i \; + \; W_{ext} \; = \; K_f \; + \; U_f \; + \; \Delta E_{th}$$

## 3) Mathematical Representation
   a.  energy conservation or work/energy
   b.  use Newton's law or momentum as needed
   c.  solve!

## 4) Assess
   a.  units?
   b.  reasonable?

# ENERGY WORKSHEET  Name _____  Problem _____

## 1) Pictorial Representation
   a. sketch of before and after
   b. coordinate system
   c. symbols for knowns and unknowns

Known:



Find:

## 2) Energy Bar Chart

What is the system? _____

Potential energies? _____

Nonconservative forces? _____

External forces? _____

Is mechanical energy conserved? _____

$$K_i \; + \; U_i \; + \; W_{ext} \; = \; K_f \; + \; U_f \; + \; \Delta E_{th}$$

## 3) Mathematical Representation
   a. energy conservation or work/energy
   b. use Newton's law or momentum as needed
   c. solve!

## 4) Assess
   a. units?
   b. reasonable?

# ENERGY WORKSHEET  Name _____  Problem _____

## 1) Pictorial Representation
    a. sketch of before and after
    b. coordinate system
    c. symbols for knowns and unknowns

Known:




Find:

## 2) Energy Bar Chart

What is the system? _____

Potential energies? _____

Nonconservative forces? _____

External forces? _____

Is mechanical energy conserved? _____

$$K_i \quad + \quad U_i \quad + \quad W_{ext} \quad = \quad K_f \quad + \quad U_f \quad + \quad \Delta E_{th}$$

## 3) Mathematical Representation
    a. energy conservation or work/energy
    b. use Newton's law or momentum as needed
    c. solve!

## 4) Assess
    a. units?
    b. reasonable?

# ENERGY WORKSHEET   Name _____   Problem _____

## 1) Pictorial Representation
  a. sketch of before and after
  b. coordinate system
  c. symbols for knowns and unknowns

```
┌─────────────────────────────────────┐
│ Known:                               │
│                                      │
│                                      │
│                                      │
│                                      │
│                                      │
│                                      │
├─────────────────────────────────────┤
│ Find:                                │
└─────────────────────────────────────┘
```

## 2) Energy Bar Chart

What is the system? _____

Potential energies? _____

Nonconservative forces? _____

External forces? _____

Is mechanical energy conserved? _____

$$K_i \quad + \quad U_i \quad + \quad W_{ext} \quad = \quad K_f \quad + \quad U_f \quad + \quad \Delta E_{th}$$

## 3) Mathematical Representation
  a. energy conservation or work/energy
  b. use Newton's law or momentum as needed
  c. solve!

## 4) Assess
  a. units?
  b. reasonable?

# ENERGY WORKSHEET  Name _____  Problem _____

## 1) Pictorial Representation
   a. sketch of before and after
   b. coordinate system
   c. symbols for knowns and unknowns

Known:



Find:

## 2) Energy Bar Chart

What is the system? _____

Potential energies? _____

Nonconservative forces? _____

External forces? _____

Is mechanical energy conserved? _____

$$K_i \quad + \quad U_i \quad + \quad W_{ext} \quad = \quad K_f \quad + \quad U_f \quad + \quad \Delta E_{th}$$

## 3) Mathematical Representation
   a. energy conservation or work/energy
   b. use Newton's law or momentum as needed
   c. solve!

## 4) Assess
   a. units?
   b. reasonable?

# ENERGY WORKSHEET  Name _____  Problem _____

## 1) Pictorial Representation
   a. sketch of before and after
   b. coordinate system
   c. symbols for knowns and unknowns

Known:



Find:

## 2) Energy Bar Chart

What is the system? _____

Potential energies? _____

Nonconservative forces? _____

External forces? _____

Is mechanical energy conserved? _____

$$K_i \quad + \quad U_i \quad + \quad W_{ext} \quad = \quad K_f \quad + \quad U_f \quad + \quad \Delta E_{th}$$

## 3) Mathematical Representation
   a. energy conservation or work/energy
   b. use Newton's law or momentum as needed
   c. solve!

## 4) Assess
   a. units?
   b. reasonable?

# ENERGY WORKSHEET   Name _____   Problem _____

## 1) Pictorial Representation
   a. sketch of before and after
   b. coordinate system
   c. symbols for knowns and unknowns

Known:

Find:

## 2) Energy Bar Chart

What is the system? _____

Potential energies? _____

Nonconservative forces? _____

External forces? _____

Is mechanical energy conserved? _____

$$K_i \quad + \quad U_i \quad + \quad W_{ext} \quad = \quad K_f \quad + \quad U_f \quad + \quad \Delta E_{th}$$

## 3) Mathematical Representation
   a. energy conservation or work/energy
   b. use Newton's law or momentum as needed
   c. solve!

## 4) Assess
   a. units?
   b. reasonable?

# ENERGY WORKSHEET   Name _____   Problem _____

## 1) Pictorial Representation
 a. sketch of before and after
 b. coordinate system
 c. symbols for knowns and unknowns

Known:




Find:

## 2) Energy Bar Chart

What is the system? _____

Potential energies? _____

Nonconservative forces? _____

External forces? _____

Is mechanical energy conserved? _____

$$K_i \quad + \quad U_i \quad + \quad W_{ext} \quad = \quad K_f \quad + \quad U_f \quad + \quad \Delta E_{th}$$

## 3) Mathematical Representation
 a. energy conservation or work/energy
 b. use Newton's law or momentum as needed
 c. solve!

## 4) Assess
 a. units?
 b. reasonable?

# ENERGY WORKSHEET  Name _____  Problem _____

## 1) Pictorial Representation
  a. sketch of before and after
  b. coordinate system
  c. symbols for knowns and unknowns

| Known: |
| --- |
| Find: |

## 2) Energy Bar Chart

What is the system? _____

Potential energies? _____

Nonconservative forces? _____

External forces? _____

Is mechanical energy conserved? _____

$$K_i \quad + \quad U_i \quad + \quad W_{ext} \quad = \quad K_f \quad + \quad U_f \quad + \quad \Delta E_{th}$$

## 3) Mathematical Representation
  a. energy conservation or work/energy
  b. use Newton's law or momentum as needed
  c. solve!

## 4) Assess
  a. units?
  b. reasonable?

# ENERGY WORKSHEET  Name _____ Problem _____

## 1) Pictorial Representation
   a. sketch of before and after
   b. coordinate system
   c. symbols for knowns and unknowns

Known:



Find:

## 2) Energy Bar Chart

What is the system? _____

Potential energies? _____

Nonconservative forces? _____

External forces? _____

Is mechanical energy conserved? _____

$+$

$0$

$-$

$$K_i \quad + \quad U_i \quad + \quad W_{ext} \quad = \quad K_f \quad + \quad U_f \quad + \quad \Delta E_{th}$$

## 3) Mathematical Representation
   a. energy conservation or work/energy
   b. use Newton's law or momentum as needed
   c. solve!

## 4) Assess
   a. units?
   b. reasonable?

# ENERGY WORKSHEET    Name _____    Problem _____

## 1) Pictorial Representation
   a. sketch of before and after
   b. coordinate system
   c. symbols for knowns and unknowns

Known:



Find:

## 2) Energy Bar Chart

What is the system? _____

Potential energies? _____

Nonconservative forces? _____

External forces? _____

Is mechanical energy conserved? _____

$$K_i \quad + \quad U_i \quad + \quad W_{ext} \quad = \quad K_f \quad + \quad U_f \quad + \quad \Delta E_{th}$$

## 3) Mathematical Representation
   a. energy conservation or work/energy
   b. use Newton's law or momentum as needed
   c. solve!

## 4) Assess
   a. units?
   b. reasonable?

# ENERGY WORKSHEET    Name _____    Problem _____

## 1) Pictorial Representation
a. sketch of before and after
b. coordinate system
c. symbols for knowns and unknowns

Known:

Find:

## 2) Energy Bar Chart

What is the system? _____

Potential energies? _____

Nonconservative forces? _____

External forces? _____

Is mechanical energy conserved? _____

$$K_i \quad + \quad U_i \quad + \quad W_{ext} \quad = \quad K_f \quad + \quad U_f \quad + \quad \Delta E_{th}$$

## 3) Mathematical Representation
a. energy conservation or work/energy
b. use Newton's law or momentum as needed
c. solve!

## 4) Assess
a. units?
b. reasonable?

# ENERGY WORKSHEET   Name _____   Problem _____

## 1) Pictorial Representation
   a. sketch of before and after
   b. coordinate system
   c. symbols for knowns and unknowns

Known:



Find:

## 2) Energy Bar Chart

What is the system? _____

Potential energies? _____

Nonconservative forces? _____

External forces? _____

Is mechanical energy conserved? _____

$$K_i \;+\; U_i \;+\; W_{ext} \;=\; K_f \;+\; U_f \;+\; \Delta E_{th}$$

## 3) Mathematical Representation
   a. energy conservation or work/energy
   b. use Newton's law or momentum as needed
   c. solve!

## 4) Assess
   a. units?
   b. reasonable?

# ENERGY WORKSHEET   Name _____   Problem _____

## 1) Pictorial Representation
    a. sketch of before and after
    b. coordinate system
    c. symbols for knowns and unknowns

Known:



Find:

## 2) Energy Bar Chart

What is the system? _____

Potential energies? _____

Nonconservative forces? _____

External forces? _____

Is mechanical energy conserved? _____

$$K_i \;+\; U_i \;+\; W_{ext} \;=\; K_f \;+\; U_f \;+\; \Delta E_{th}$$

## 3) Mathematical Representation
    a. energy conservation or work/energy
    b. use Newton's law or momentum as needed
    c. solve!

## 4) Assess
    a. units?
    b. reasonable?

# ENERGY WORKSHEET  Name _____  Problem _____

## 1) Pictorial Representation

    a. sketch of before and after

    b. coordinate system

    c. symbols for knowns and unknowns

Known:



Find:

## 2) Energy Bar Chart

What is the system? _____

Potential energies? _____

Nonconservative forces? _____

External forces? _____

Is mechanical energy conserved? _____

$$K_i + U_i + W_{ext} = K_f + U_f + \Delta E_{th}$$

## 3) Mathematical Representation

    a. energy conservation or work/energy

    b. use Newton's law or momentum as needed

    c. solve!

## 4) Assess

    a. units?

    b. reasonable?

# ENERGY WORKSHEET  Name _____  Problem _____

## 1) Pictorial Representation
    a. sketch of before and after
    b. coordinate system
    c. symbols for knowns and unknowns

Known:



Find:

## 2) Energy Bar Chart

What is the system? _____

Potential energies? _____

Nonconservative forces? _____

External forces? _____

Is mechanical energy conserved? _____

$$K_i \;+\; U_i \;+\; W_{ext} \;=\; K_f \;+\; U_f \;+\; \Delta E_{th}$$

## 3) Mathematical Representation
    a. energy conservation or work/energy
    b. use Newton's law or momentum as needed
    c. solve!

## 4) Assess
    a. units?
    b. reasonable?

# ENERGY WORKSHEET   Name _____   Problem _____

## 1) Pictorial Representation
    a. sketch of before and after
    b. coordinate system
    c. symbols for knowns and unknowns

Known:



Find:

## 2) Energy Bar Chart

What is the system? _____

Potential energies? _____

Nonconservative forces? _____

External forces? _____

Is mechanical energy conserved? _____

$$K_i \;+\; U_i \;+\; W_{ext} \;=\; K_f \;+\; U_f \;+\; \Delta E_{th}$$

## 3) Mathematical Representation
    a. energy conservation or work/energy
    b. use Newton's law or momentum as needed
    c. solve!

## 4) Assess
    a. units?
    b. reasonable?